Differential Geometry of Frame Bundles

Mathematics and Its Applications

Differential Geometry of Frame Bundles

Luis A. Cordero

Department of Geometry and Topology, University of Santiago de Compostela, Spain

C. T. J. Dodson

Department of Mathematics, University of Lancaster, United Kingdom

and

Manuel de León

C.E.C.I.M.E.-C.S.I.C., Madrid, Spain

KLUWER ACADEMIC PUBLISHERS

DORDRECHT / BOSTON / LONDON

Library of Congress Cataloging in Publication Data

Cordero, L. A.
Differential geometry of frame bundles / Luis A. Cordero, C.T.J. Dodson, and Manuel de León.
p. cm.
Bibliography: p.
Includes index.
ISBN 0-7923-0012-2
1. Frame bundles. 2. Jets (Topology) 3. Differentiable manifolds. I. Dodson, C. T. J. II. León, Manuel De, 1953- III. Title.
QA612.63.C67 1988
514'.224--dc19 88-29778
CIP

ISBN 0-7923-0012-2

Published by Kluwer Academic Publishers,
P.O. Box 17, 3300 AA Dordrecht, The Netherlands.

Kluwer Academic Publishers incorporates
the publishing programmes of
D. Reidel, Martinus Nijhoff, Dr W. Junk and MTP Press.

Sold and distributed in the U.S.A. and Canada
by Kluwer Academic Publishers,
101 Philip Drive, Norwell, MA 02061, U.S.A.

In all other countries, sold and distributed
by Kluwer Academic Publishers Group,
P.O. Box 322, 3300 AH Dordrecht, The Netherlands.

SERIES EDITOR'S PREFACE

Approach your problems from the right end and begin with the answers. Then one day, perhaps you will find the final question.

'The Hermit Clad in Crane Feathers' in R. van Gulik's *The Chinese Maze Murders*.

It isn't that they can't see the solution. It is that they can't see the problem.

G.K. Chesterton. *The Scandal of Father Brown* 'The point of a Pin'.

Growing specialization and diversification have brought a host of monographs and textbooks on increasingly specialized topics. However, the "tree" of knowledge of mathematics and related fields does not grow only by putting forth new branches. It also happens, quite often in fact, that branches which were thought to be completely disparate are suddenly seen to be related.

Further, the kind and level of sophistication of mathematics applied in various sciences has changed drastically in recent years: measure theory is used (non-trivially) in regional and theoretical economics; algebraic geometry interacts with physics; the Minkowsky lemma, coding theory and the structure of water meet one another in packing and covering theory; quantum fields, crystal defects and mathematical programming profit from homotopy theory; Lie algebras are relevant to filtering; and prediction and electrical engineering can use Stein spaces. And in addition to this there are such new emerging subdisciplines as "experimental mathematics", "CFD", "completely integrable systems", "chaos, synergetics and large-scale order", which are almost impossible to fit into the existing classification schemes. They draw upon widely different sections of mathematics. This programme, Mathematics and Its Applications, is devoted to new emerging (sub)disciplines and to such (new) interrelations as exempla gratia:

- a central concept which plays an important role in several different mathematical and/or scientific specialized areas;
- new applications of the results and ideas from one area of scientific endeavour into another;
- influences which the results, problems and concepts of one field of enquiry have and have had on the development of another.

The Mathematics and Its Applications programme tries to make available a careful selection of books which fit the philosophy outlined above. With such books, which are stimulating rather than definitive, intriguing rather than encyclopaedic, we hope to contribute something towards better communication among the practitioners in diversified fields.

It is, in fact, quite amazing just how important bundles of all kinds have become in physics and in mathematics. On the one hand, gauge theories have a lot to do with that; after all, a particle field is simply a section of a vector bundle, and a Yang-Mills potential is a connection one-form. On the other hand, there is the important and natural trend in mathematics to work globally and coordinate free, i.e. to do (also) global (nonlinear) analysis. The natural vehicle for that are bundles; with frame bundles as the natural carriers for the more important structures such as connections (which is 'global' for differential equation).

Just for fun I punched up the STN database (invaluable for this sort of thing) and checked how many articles appeared in the physics literature in 1986 with the world 'bundle' occurring in the title or abstract. The answer was a solid 136; the mathematics base scored 438. If one wants all this stuff to be accessible one needs to have (at least in part) the sort of background knowledge contained in this volume. And that is why I am happy to welcome this clear exposé of frame-bundles-and-what-they-are-good-for in this series.

Incidentally, geometry as the language of physics, is nothing much new, though the kind and level of abstraction have changed drastically:

"Those few things having been considered,the whole matter is reduced to pure geometry, which is the one aim of physics and mechanics."

G.W. Leibniz

The unreasonable effectiveness of mathematics in science ...

Eugene Wigner

Well, if you know of a better 'ole, go to it.

Bruce Bairnsfather

What is now proved was once only imagined.

William Blake

As long as algebra and geometry proceeded along separate paths, their advance was slow and their applications limited.

But when these sciences joined company they drew from each other fresh vitality and thenceforward marched on at a rapid pace towards perfection.

Joseph Louis Lagrange.

Bussum, October 1988

Michiel Hazewinkel

Contents

Preface

The study of frame bundles can be motivated by the intrinsic richness of their geometry, by their exhibition of a fascinating interplay between algebra and geometry, and by their applications.

The frame bundle and its subbundles for a given manifold provide an organization of possible geometries that can be supported, and encode global topological features. Viewed another way, these bundles exemplify the available trivial and twisted smooth products of linear groups over the manifold, so forming an important class of principal bundles. They provide the natural context in which to elaborate the theory of linear connections, parallel transport of frames of references being intuitively appealing through common experience, yet representing subtle facets of natural philosophy for general relativity theory.

One of the main themes of this book is to describe how and with what effect various structures on the base manifold admit lifts to the frame bundle, and to the second frame bundle. For example, we study the prolongation of G–structures and fields, giving prominence to the lifting of metrics and connections with their concomitants. We see also how the superstructure embodied in the notion of a system of connections allows a simultaneous grasp of all possible connections as pullbacks of a canonical universal connection.

Our approach is via jets and we give an intrinsic, coordinate–free development, but we provide the detailed formulae needed for computations in any particular cases. As prerequisites, we suppose that the reader has taken a first course in differential geometry, has met the ideas of metrics, connections and curvature forms, and has some familiarity with the basics of Lie groups and their actions. Also, some terminology from category theory is used for convenience but in no case would this obscure the development for those with little experience of it.

Throughout we shall work in smooth categories.

Chapter 1

The Functor J_p^1

Introduction

A natural and useful equivalence relation for real–valued maps on Euclidean $\mathbf{R}^p$ is that of having the same gradient vector at the origin. For maps on $\mathbf{R}^p$ taking values in a smooth n–manifold M, we call them 1–jet equivalent if their composite with real–valued maps on M always yields the same gradient vector at the origin of $\mathbf{R}^p$. Thus, we factor through M the gradient classes of real–valued maps on $\mathbf{R}^p$.

This process can be made functorial and the objects so generated are manifolds; in particular, for $p = 1$ we recover the tangent bundle and for $p = n$ we recover the frame bundle. Moreover, if M is a Lie group, then our functor generates a new Lie group and similarly for Lie algebras, correctly locking together the two processes. The functor even preserves Lie group actions, effectivity and transitivity.

1.1 The Bundle $J_p^1 M \longrightarrow M$

Let $C^\infty(\mathbf{R}^p)$ be the algebra of C^∞–functions on the Euclidean space $\mathbf{R}^p$ with natural coordinates $(t^1, \ldots, t^p)$. Let $f, g \in C^\infty(\mathbf{R}^p)$; we say that f is equivalent to g if $f(0) = g(0)$ and $(\partial/\partial t^\alpha)(f) = (\partial/\partial t^\alpha)(g)$ at $t = 0 \in \mathbf{R}^p$ for every $\alpha = 1, 2, ..., p$; clearly, this is an equivalence relation.

Now, let M be an n–dimensional manifold. Consider the set $S_p(M)$ of all maps $\phi: \mathbf{R}^p \longrightarrow M$, and take two elements $\phi, \psi \in S_p(M)$; we say that ϕ is equivalent to ψ if $(f \circ \phi)$ is equivalent to $(f \circ \psi)$ for every $f \in C^\infty(M)$. This is again an equivalence relation; we denote by $j^1(\phi)$ the equivalence class of $\phi \in S_p(M)$, and we shall call it a 1–*jet* in M at $\phi(0)$. Let $J_p^1 M$ denote the set of all equivalence classes in $S_p(M)$.

In order to introduce a manifold structure in J^1_pM, we define local charts on it as follows: let $(U, x^1, \ldots, x^n)$ be a local chart in M; define coordinate functions

$$\{x^i, x^i_\alpha; i = 1, 2, \ldots, n,\ \alpha = 1, 2, \ldots, p\}$$

on J^1_pU by

$$\begin{aligned} x^i(j^1(\phi)) &= x^i(\phi(0)) , \\ x^i_\alpha(j^1(\phi)) &= \frac{\partial(x^i \circ \phi)}{\partial t^\alpha}_{|0} , \end{aligned}$$

for every $j^1(\phi) \in J^1_pU$.

It is straightforward to see that J^1_pM becomes an $(n + pn)$–manifold; indeed if (V, y^i) is another local chart on M, with $U \cap V \neq \emptyset$, and if $(J^1_pV, y^i, y^i_\alpha)$ is the local chart induced on J^1_pM, then the change of coordinates is given by

$$\text{(1.1)} \qquad \left\{ \begin{aligned} y^i &= y^i(x^1, \ldots, x^n) , \\ y^i_\alpha &= \frac{\partial y^i}{\partial x^k} x^k_\alpha , \end{aligned} \right.$$

in $J^1_pU \cap J^1_pV$, and hence the Jacobian matrix of (1.1) is:

$$\text{(1.2)} \qquad \begin{pmatrix} \left(\frac{\partial y^i}{\partial x^j}\right) & 0 & \ldots & 0 \\ \left(\frac{\partial^2 y^i}{\partial x^j \partial x^k} x^k_1\right) & \left(\frac{\partial y^i}{\partial x^j}\right) & \ldots & 0 \\ \vdots & \vdots & \ddots & \vdots \\ \left(\frac{\partial^2 y^i}{\partial x^j \partial x^k} x^k_p\right) & 0 & \ldots & \left(\frac{\partial y^i}{\partial x^j}\right) \end{pmatrix} .$$

Let $\pi: J^1_pM \longrightarrow M$ be the canonical projection map given by $\pi(j^1(\phi)) = \phi(0)$. It is obvious that π is a submersion.

Remark 1.1.1 Sometimes it will be useful to write the induced coordinate functions on J^1_pU as

$$\{x^A; A = 1, 2, \ldots, n + pn\} \quad , \text{ with } \quad x^{\alpha n + j} = x^j_\alpha .$$

There is an alternative way to describe the points in the differentiable manifold J^1_pM.

Let $j^1(\phi) \in J^1_pM$ be an arbitrary point, and let $\phi^\alpha: \mathbf{R} \longrightarrow M$ be the differentiable curve given by $\phi^\alpha(t) = (0, \ldots, t, \ldots, 0)$, with t at the α–th place; then, associated to $j^1(\phi)$ there is a unique $(p+1)$–tuple $[x; X_1, \ldots, X_p]$ given by

$$x = \phi(0) \quad , \quad X_\alpha = (\phi^\alpha)_* \left(\frac{d}{dt}_{|0}\right) ,$$

where d/dt is the canonical vector field tangent to $\mathbf{R}$. Clearly

$$j^1(\phi) \longmapsto [x; X_1, \ldots, X_p]$$

is well defined and bijective.

From now on, we shall write $[x; X_1, \ldots, X_p]$ simply as $[x; X_\alpha]$ and shall identify $j^1(\phi) \equiv [x; X_\alpha]$ if there is no confusion.

The correspondence $J_p^1 : M \longrightarrow J_p^1 M$ has the following functorial properties, among many others to be described later:

Functorial properties of J_p^1

Let $h: M \longrightarrow N$ be differentiable. Then h induces a canonical differentiable map

$$h^1 : J_p^1 M \longrightarrow J_p^1 N$$

given by

$$h^1(j^1(\phi)) = j^1(h \circ \phi) , \quad \text{for any } j^1(\phi) \in J_p^1 M ,$$

and, in terms of the previous identification, we have:

$$h^1([x; X_\alpha]) = [h(x); h_* X_\alpha] .$$

In terms of the induced charts, h^1 is given as follows: let (U, x^i) be a local chart on M, $x \in U$, and (V, y^a) a local chart on N, $h(x) \in V$, $1 \leq a \leq dim\ N$; assume that h is locally given with respect to these local systems as:

$$h: \quad y^a = h^a(x^1, \ldots, x^n) ;$$

then, with respect to the induced charts $(J_p^1 U, x^i, x^i_\alpha)$, $(J_p^1 V, y^a, y^a_\alpha)$, the induced map h^1 is given by

$$h^1: \quad y^a = h^a(x^1, \ldots, x^n) , \ y^a_\alpha = \frac{\partial h^a}{\partial x^j} x^j_\alpha .$$

Obviously $h \circ \pi = \pi \circ h^1$.

Now, it is straightforward to compute h^1_*, which is given by

$$h^1_* \equiv \begin{pmatrix} \left(\frac{\partial h^a}{\partial x^j}\right) & 0 & \ldots & 0 \\ \left(\frac{\partial^2 h^a}{\partial x^j \partial x^k} x^k_1\right) & \left(\frac{\partial h^a}{\partial x^j}\right) & \ldots & 0 \\ \vdots & \vdots & \ddots & \vdots \\ \left(\frac{\partial^2 h^a}{\partial x^j \partial x^k} x^k_p\right) & 0 & \ldots & \left(\frac{\partial h^a}{\partial x^j}\right) \end{pmatrix} .$$

It is clear that if $h: M \longrightarrow N$ is a diffeomorphism, then the induced map $h^1: J^1_pM \longrightarrow J^1_pN$ is also a diffeomorphism and, moreover, $(h^1)^{-1} = (h^{-1})^1$.

Also, if $f : M \longrightarrow M'$ and $g : M' \longrightarrow M''$ are differentiable maps, then $(g \circ f)^1 = g^1 \circ f^1$, and if $1_M : M \longrightarrow M$ denotes the identity map, then $(1_M)^1 = 1_{J^1_pM}$.

Let M_1, M_2 be differentiable manifolds, $M_1 \times M_2$ the product manifold. Then $J^1_p(M_1 \times M_2)$ and $J^1_pM_1 \times J^1_pM_2$ are diffeomorphic. Define

$$\phi: J^1_p(M_1 \times M_2) \longrightarrow J^1_pM_1 \times J^1_pM_2$$

by setting

$$\phi(j^1(f)) = (j^1(f_1), j^1(f_2)) \quad , \quad j^1(f) \in J^1_p(M_1 \times M_2) \ ,$$

where $f_i = \rho_i \circ f$, and $\rho_i: M_1 \times M_2 \longrightarrow M_i$ the canonical projection.

Conversely, define $\psi: J^1_pM_1 \times J^1_pM_2 \longrightarrow J^1_p(M_1 \times M_2)$ by

$$\psi(j^1(f_1), j^1(f_2)) = j^1(f) \ , \quad j^1(f_i) \in J^1_pM_i \ , \quad i = 1,2 \ ,$$

where $f = (f_1, f_2): \mathbb{R}^p \longrightarrow M_1 \times M_2$ is given by $f(x) = (f_1(x), f_2(x))$.

It is clear that $\psi = \phi^{-1}$ and we shall always identify

$$J^1_p(M_1 \times M_2) \stackrel{\simeq}{\longleftrightarrow} J^1_pM_1 \times J^1_pM_2 \ .$$

Moreover, if $\mu : M_1 \times M_2 \longrightarrow N$ is a differentiable map, then the map $\mu^1 \circ \psi : J^1_pM_1 \times J^1_pM_2 \longrightarrow J^1_pN$ will be denoted simply as μ^1; analogously, if $h_i : M_i \longrightarrow N_i, i = 1,2$, then the induced maps $(h_1 \times h_2)^1$ and $h^1_1 \times h^1_2$ also will be identified.

Canonical lifts of vector fields to J^1_pM

Let X be a vector field over M, and let $\{\phi_t\}$ be the 1–parameter group of local transformations of M induced by X; then $\{\phi^1_t\}$ is again a 1–parameter group of local transformations of J^1_pM, and hence it defines a unique vector field over J^1_pM which will be denoted by X^C.

If X is given by

$$X = X^i \frac{\partial}{\partial x^i} \ ,$$

with respect to a local chart (U, x^i) on M, then one easily deduces that X^C is given, with respect to the induced chart $(J^1_pU, x^i, x^i_\alpha)$, by

$$X^C = X^i \frac{\partial}{\partial x^i} + x^j_\alpha \frac{\partial X^i}{\partial x^j} \frac{\partial}{\partial x^i_\alpha} \ . \tag{1.3}$$

We shall say that X^C is *the complete lift of* X *to* J^1_pM.

On the other hand, for each $\alpha = 1, 2, \dots, p$, there exists on J^1_pM a vector field $X^{(\alpha)}$, uniquely associated to X, defined as follows: with respect to $(J^1_pU, x^i, x^i_\alpha)$

$$X^{(\alpha)} = X^i \frac{\partial}{\partial x^i_\alpha} . \tag{1.4}$$

The vector field $X^{(\alpha)}$ will be called *the* α*–th vertical lift of* X *to* J^1_pM.

The following are important properties of these lifts:

(1): Since, locally,

$$\left(\frac{\partial}{\partial x^i}\right)^C = \frac{\partial}{\partial x^i} , \quad \left(\frac{\partial}{\partial x^i}\right)^{(\alpha)} = \frac{\partial}{\partial x^i_\alpha} ,$$

it follows that every differential form over J^1_pM is completely determined by its action over the lifts X^C and $X^{(\alpha)}$ of every vector field X over M.

(2): A direct computation from (1.3) and (1.4) shows the following identities: for any X, Y over M and every $\alpha, \beta \in \{1, 2, \dots, p\}$

$$\begin{aligned} [X^C, Y^C] &= [X, Y]^C , \\ [X^C, Y^{(\alpha)}] &= [X, Y]^{(\alpha)} , \\ [X^{(\alpha)}, Y^{(\beta)}] &= 0 . \end{aligned}$$

An alternative definition of these lifts will be given in Chapter 3.

Two particular cases

Two particular values for p, namely $p = 1$ and $p = n = dim\ M$, lead to two specially important situations:

<u>Assume $p = 1$</u>

Then $\pi: J^1_pM \longrightarrow M$ is precisely the tangent bundle of M, and we shall denote it by $\pi_M: TM \longrightarrow M$. Moreover, if (U, x^i) is a local chart on M, we shall denote by $(TU, x^i, \dot{x}^i)$ the induced chart on TM.

Observe that the vector bundle structure on TM is locally given as follows: let X, Y be vectors tangent at $x = (x^1, \dots, x^n) \in U$ with coordinates $X = (x^i, \dot{x}^i)$, $Y = (x^i, \dot{y}^i)$, then $X + Y = (x^i, \dot{x}^i + \dot{y}^i)$.

It is easy to check that, for any vector field X on M, X^C is just the complete lift of X to TM in the usual sense of the theory of lifts to the tangent bundle [88]. Also, the vertical lift $X^{(1)}$ is just the so called vertical lift of X to TM.

Finally, if $f: M \longrightarrow N$ is a differentiable map, then we shall denote the induced map by $Tf: TM \longrightarrow TN$, which is the linear tangent map to f in the usual sense. If $\phi: \mathbf{R} \longrightarrow M$ is differentiable, then $j^1(\phi)$ will be denoted as $\dot{\phi}$.

Assume $p = n = \dim M$

Then the total space FM of the frame bundle of M is an open submanifold of $J_n^1 M$, because FM can be considered as the set of 1-jets at $0 \in \mathbf{R}^n$ of local diffeomorphisms of open neighbourhoods of $0 \in \mathbf{R}^n$ into M.

In fact, the differentiable manifold structure defined over $J_n^1 M$ induces on the open submanifold FM its usual structure with respect to which

$$\pi_M: FM \longrightarrow M$$

is a principal fibre bundle with structure group $Gl(n, \mathbf{R})$.

If (U, x^i) is a local chart on M, then $(J_n^1 U, x^i, x_\alpha^i)$ induces on FM a local chart (FU, x^i, x_α^i), where we put $FU = J_n^1 U \cap FM$.

It is worthwhile to remark that (x_α^i) is a non-singular $n \times n$ matrix of functions over FU whose inverse will be denoted (x_i^α).

Diffeomorphisms $\alpha_M^{p,1}$ and $\alpha_M^{1,p}$

Specializing some general results of A. Morimoto [68], we can state:

Theorem 1.1.2 *There exist canonical mutually inverse diffeomorphisms*

$$\begin{aligned} \alpha_M^{p,1} &: TJ_p^1 M \longrightarrow J_p^1 TM \\ \alpha_M^{1,p} &: J_p^1 TM \longrightarrow TJ_p^1 M \end{aligned}$$

Proof. For their further use, we shall recall the definitions of both maps. Details of the proof, in a more general context, can be found in [68].

Let $\dot{\phi} \in T_{\phi(0)} J_p^1 M$ be the tangent vector to $\phi: \mathbf{R} \longrightarrow J_p^1 M$ at $\phi(0)$; then, there exist $\psi: \mathbf{R} \times \mathbf{R}^p \longrightarrow M$ and $\delta > 0$ such that $\phi(t) = j^1(\psi_t)$ for $|t| < \delta$, where $\psi_t(u) = \psi^u(t) = \psi(t, u), t \in \mathbf{R}, u \in \mathbf{R}^p$.

Define $\Psi: \mathbf{R}^p \longrightarrow TM$ by $\Psi(u) = \dot{\psi}^u$; then $\alpha_M^{p,1}(\dot{\phi}) = j^1(\Psi)$.

Analogously, let $\eta: \mathbf{R}^p \longrightarrow TM$ be a differentiable map; then, there exist $\phi: \mathbf{R}^p \times \mathbf{R} \longrightarrow TM$ and $\delta > 0$ such that $\eta(t) = \dot{\phi}_t$ for $|t| < \delta$, where $\phi_t(u) = \phi^u(t) = \phi(t, u), t \in \mathbf{R}^p, u \in \mathbf{R}$.

Define $\Phi: \mathbf{R} \longrightarrow J_p^1 M$ by $\Phi(u) = j^1(\phi^u)$; then $\alpha_M^{1,p}(j^1(\eta)) = \dot{\Phi}$.

Locally, the map $\alpha_M^{p,1}$ is given as follows: let (U, x^i) be a chart on M, and let $(TJ_p^1 U, x^i, x_\alpha^i, \dot{x}^i, \dot{x}_\alpha^i)$, $(J_p^1 TU, y^i, \dot{y}^i, (y^i)_\alpha, (\dot{y}^i)_\alpha)$ be the induced charts on $TJ_p^1 U$ and $J_p^1 TU$, respectively; then,

$$\alpha_M^{p,1}: \; y^i = x^i, \; \dot{y}^i = \dot{x}^i, \; (y^i)_\alpha = x_\alpha^i, \; (\dot{y}^i)_\alpha = \dot{x}_\alpha^i \; . \tag{1.5}$$

The local expression for $\alpha^{1,p}_M$ is obvious.□

Moreover, for any differentiable $f: M \longrightarrow N$

$$\begin{aligned}(Tf)^1 \circ \alpha^{1,p}_M &= \alpha^{1,p}_M \circ Tf^1 \\ Tf^1 \circ \alpha^{1,p}_M &= \alpha^{p,1}_M \circ (Tf)^1 .\end{aligned}$$

In particular, $\alpha^{1,p}_M$ can be used to obtain an alternative description of the complete lift X^C of a vector field X over M as follows.

Let us consider X as a section $X: M \longrightarrow TM$ of the tangent bundle, i.e. $\pi_M \circ X = 1_M$. Then the composition

$$\alpha^{1,p}_M \circ X^1: J^1_pM \longrightarrow TJ^1_pM$$

is again a section, and hence it defines a vector field over J^1_pM; it is easy to check that $X^C = \alpha^{1,p}_M \circ X^1$.

1.2 J^1_pG for a Lie group G

Let G be a Lie group with Lie algebra $\mathfrak{g}$. Then J^1_pG inherits a canonical Lie group structure.

Let $\mu: G \times G \longrightarrow G$ be the group multiplication; then, the induced map $\mu^1: J^1_pG \times J^1_pG \longrightarrow J^1_pG$ defines a Lie group multiplication on J^1_pG which is compatible with the manifold structure of J^1_pG. In fact, for all $j^1(\phi), j^1(\psi) \in J^1_pG$,

$$j^1(\phi) \cdot j^1(\psi) = j^1(\phi\psi) \ , \quad \{j^1(\phi)\}^{-1} = j^1(\phi^{-1}) \ ,$$

where $(\phi\psi), \phi^{-1}: \mathbf{R}^p \longrightarrow G$ are differentiable maps given by

$$(\phi\psi)(x) = \phi(x)\psi(x) \ , \quad \phi^{-1}(x) = \{\phi(x)\}^{-1} \ .$$

The unit element e^1 of J^1_pG is the 1–jet at $0 \in \mathbf{R}^p$ of the constant map sending $\mathbf{R}^p$ into the unit element e of G.

This Lie group structure on J^1_pG can be equivalently described as follows: put $j^1(\phi) \equiv [a; X_\alpha]$, $j^1(\psi) \equiv [b; Y_\alpha]$; then $j^1(\phi\psi) \equiv [ab; Z_\alpha]$ where Z_α is the tangent vector at $t = 0$ of the curve $(\phi\psi)^\alpha: \mathbf{R} \longrightarrow G$ defined by $(\phi\psi)^\alpha = \mu \circ (\phi^\alpha, \psi^\alpha)$ with $(\phi^\alpha, \psi^\alpha): \mathbf{R} \longrightarrow G \times G$ given by $(\phi^\alpha, \psi^\alpha)(t) = (\phi^\alpha(t), \psi^\alpha(t))$, for each $\alpha = 1, 2, \ldots, p$. Therefore,

$$Z_\alpha = (\phi\psi)^\alpha_* \left(\frac{d}{dt}_{|0} \right) = \mu_*(X_\alpha, Y_\alpha) = (R_b)_* X_\alpha + (L_a)_* Y_\alpha \ ,$$

and then we can write

$$[a; X_\alpha] \cdot [b; Y_\alpha] = [ab; (R_b)_* X_\alpha + (L_a)_* Y_\alpha] \; . \tag{1.6}$$

Obviously, $\pi: J^1_p G \longrightarrow G$ is a Lie group homomorphism.

Now, take an arbitrary point $[a; X_\alpha] \in J^1_p G$, and denote by $B_\alpha \in \mathfrak{g}$ the unique element such that $X_\alpha = (R_a)_* B_\alpha, \alpha = 1, 2, \ldots, p$. Then, from (1.6), one easily deduces that the injection

$$\begin{array}{rcl} J^1_p G & \longrightarrow & G \times (\times_p \mathfrak{g}) \\ {[a; X_\alpha]} & \longmapsto & [a; B_\alpha] \end{array} \tag{1.7}$$

satisfies

$$[a; X_\alpha] \cdot [b; Y_\alpha] \longmapsto [ab; B_\alpha + ad\; a^{-1} B'_\alpha] \; , \tag{1.8}$$

where $\times_p \mathfrak{g} = \mathfrak{g} \times \overset{p}{\cdots} \times \mathfrak{g}$ and $Y_\alpha = (R_b)_* B'_\alpha, B'_\alpha \in \mathfrak{g}$.

Denote by $G \times_{ad} (\times_p \mathfrak{g})$ the semidirect product of G and the Abelian Lie group $\times_p \mathfrak{g}$ via the canonical extension of the adjoint representation of G, $ad: G \longrightarrow Aut\,(\mathfrak{g})$, and by $\mathfrak{g} \times_{Ad} (\times_p \mathfrak{g})$ the semidirect product of the Lie algebra $\mathfrak{g}$ and the Abelian Lie algebra $\times_p \mathfrak{g}$; then from (1.7) and (1.8) it follows:

Theorem 1.2.1 *There exists a canonical isomorphism of Lie groups*

$$J^1_p G \xrightarrow{\simeq} G \times_{ad} (\times_p \mathfrak{g})$$

given by (1.7) *and, in consequence (see* Corollary 1.3.2*), a canonical isomorphism of Lie algebras*

$$J^1_p \mathfrak{g} \xrightarrow{\simeq} \mathfrak{g} \times_{Ad} (\times_p \mathfrak{g})$$

$J^1_p G$ acting on $J^1_p M$

Next we consider a Lie group action

$$\begin{array}{rcl} \rho: G \times M & \longrightarrow & M \\ (a, x) & \longmapsto & a \cdot x \end{array}$$

with induced maps

$$\begin{array}{rclcrcl} \rho_x: G & \longrightarrow & M & : & a & \longmapsto & a \cdot x \\ \tau_a: M & \longrightarrow & M & : & x & \longmapsto & a \cdot x \end{array}$$

Then J^1_pG acts on J^1_pM by

$$\begin{aligned} \rho^1\colon J^1_pG \times J^1_pM &\longrightarrow J^1_pM \\ (j^1(\phi), j^1(f)) &\longmapsto j^1(\phi f) , \end{aligned}$$

where $(\phi f)(t) = \rho(\phi(t), f(t)) = \phi(t) \cdot f(t)$, $t \in \mathbf{R}^p$.

In fact, if we put $j^1(\phi) \equiv [a; B_\alpha]$, $B_\alpha \in \mathfrak{g}$ (by virtue of (1.7)) and $j^1(f) \equiv [x; X_\alpha]$, $X_\alpha \in T_xM, \alpha = 1, 2, ..., p$, then $j^1(\phi f) \equiv [a \cdot x; Y_\alpha]$ where $Y_\alpha \in T_xM$ is given, for each α, by

$$Y_\alpha = (\rho_x \circ R_a)_* B_\alpha + (\tau_a)_* X_\alpha = (B^*_\alpha)_{a \cdot x} + (\tau_a)_* X_\alpha ,$$

where B^*_α stands for the vector field on M induced by $B_\alpha \in \mathfrak{g}$. Thus, the action of J^1_pG on J^1_pM can be written as follows:

$$[a; B_\alpha] \cdot [x; X_\alpha] = [a \cdot x; (B^*_\alpha)_{a \cdot x} + (\tau_a)_* X_\alpha] . \tag{1.9}$$

From this formula one easily deduces:

Proposition 1.2.2 *If G acts effectively on M then J^1_pG does so on J^1_pM.*

Also, from (1.9) we find:

$$\begin{aligned} [a; 0] \cdot [x; X_\alpha] &= [a \cdot x; (\tau_a)_* X_\alpha] , \\ [e; B_\alpha] \cdot [x; X_\alpha] &= [x; (B^*_\alpha)_x + X_\alpha] . \end{aligned}$$

Proposition 1.2.3 *If G acts transitively on M then J^1_pG does so on J^1_pM.*

Proof. Let $j^1(f), j^1(g) \in J^1_pM$; then, for any $t \in dom(f) \cap dom(g)$, there exists a unique element $a_t \in G$ such that $f(t) = a_t \cdot g(t)$. Thus we obtain a map $h\colon U \subset \mathbf{R}^p \to G$ given by $h(t) = a_t$.

Note that h satisfies $f = \rho \circ (h \times g) \circ \Delta$, $\Delta\colon \mathbf{R}^p \to \mathbf{R}^p \times \mathbf{R}^p$ denoting the diagonal injection, and therefore $j^1(f) \in J^1_pG$ only depends on $j^1(f)$ and $j^1(g)$. Moreover, by construction, $j^1(f) = j^1(h) \cdot j^1(g)$, which ends the proof.□

Proposition 1.2.4 *Let $H = \{a \in G \mid a \cdot x = x\}$ be the isotropy group of $x \in M$ with respect to $\rho\colon G \times M \longrightarrow M$; then, with respect to the induced action, J^1_pH is the isotropy group of any arbitrary $y \in J^1_pM$, with $y = j^1(f)$ and $f(\mathbf{R}^p) = x$.*

Proof. Let $j^1(f) \in H$; then $j^1(h) \cdot y = j^1(hf)$. But $(hf)(t) = h(t) \cdot f(t) = h(t) \cdot x = x$, hence $hf = f$ and $j^1(h) \cdot y = y$.

Conversely, let $j^1(h) \in J^1_pG$ be an element of the isotropy group of y; let us denote $j^1(h) \equiv [a; B_\alpha]$, $B_\alpha \in g$, $x \equiv [x; 0]$. Then, by (1.9), $[a; B_\alpha] \cdot [x; 0] =$

$[a\cdot x; (B^*_\alpha)_{a\cdot x}]$, from where we deduce, on the one hand, that $a\cdot x = x$, hence $a \in H$, and on the other hand $(B^*_\alpha)_x = 0$. But $(B^*_\alpha)_x$ is the vector tangent to the curve $(exp\ sB_\alpha)\cdot x$ on M at the point x; therefore the curve reduces to the point x, which implies that $(exp\ sB_\alpha) \in H$ for every s, and hence $B_\alpha \in \mathfrak{h}$, the Lie algebra of H. Thus $[a; B_\alpha] \in J^1_p H$.□

Corollary 1.2.5 *Let $H \subset G$ be a closed subgroup. Then $J^1_p(G/H)$ is a homogeneous space; in fact $J^1_p(G/H) \simeq J^1_p G/J^1_p H$.*

1.3 $J^1_p V$ for a vector space V

Let V be a (real) vector space, $dim\ V = m$. Fix, once for all, a basis $\{e_a, 1 \leq a \leq m\}$ of V, and consider V as an m-dimensional manifold. Then $J^1_p V$ inherits a vector space structure: for any $j^1(f), j^1(g) \in J^1_p V$ and $\lambda \in \mathbf{R}$, define

$$j^1(f) + j^1(g) = j^1(f+g)\ ,\quad \lambda(j^1(f)) = j^1(\lambda f)\ ,$$

where $f+g, \lambda f\colon \mathbf{R}^p \longrightarrow \mathbf{R}$ are defined in the usual way.

An induced basis of $J^1_p V$ is constructed as follows. We have canonical maps $f_a, f_{a_\alpha}\colon \mathbf{R}^p \longrightarrow V$ given by:

$$f_a(t) = e_a\ ,\quad f_{a_\alpha}(t) = t^\alpha e_a\ ,\quad t = (t^1, \ldots, t^p) \in \mathbf{R}^p\ ,$$

and set $E_a = j^1(f_a)$, $E_{a_\alpha} = j^1(f_{a_\alpha})$; then, $\{E_a, E_{a_\alpha}\}$ is a basis for $J^1_p V$.

On the other hand, the vector space $V^{1+p} = V \times V^p$ possesses a canonical basis $\{E'_a, E'_{a_\alpha}\}$ induced from $\{e_a\}$ given by:

$$E'_a = (e_a, 0, \ldots, 0)\ ,$$

and

$$E'_{a_\alpha} = (0, \ldots, e_a, \ldots, 0) \text{ with } e_a \text{ at the } (\alpha+1)\text{–place}\ ,$$

and the correspondence $E_a \longrightarrow E'_a$, $E_{a_\alpha} \longrightarrow E'_{a_\alpha}$ defines an isomorphism of vector spaces $J^1_p V$ and V^{1+p}; this isomorphism being, in fact, a diffeomorphism, hereafter $J^1_p V$ and V^{1+p} will be identified without explicit mention.

For later use, let us specialize this isomorphism of vector spaces for $V = \mathbf{R}^n$. If $[x; X_\alpha] \in J^1_p \mathbf{R}^n$ is the point with coordinates (x^h, X^h_α), i.e. $x = (x^1, \ldots, x^n)$, $X_\alpha = X^i_\alpha(\partial/\partial x^i)_x$, then the canonical isomorphism $J^1_p \mathbf{R}^n \simeq \mathbf{R}^{n+np}$ is expressed by

$$[x; X_\alpha] \in J^1_p \mathbf{R}^n \longmapsto \begin{pmatrix} x^h \\ X^h_\alpha \end{pmatrix} \in \mathbf{R}^{n+pn}\ . \tag{1.10}$$

$J^1_p\mathfrak{g}$ for a Lie algebra $\mathfrak{g}$

Consider the case that $V = \mathfrak{g}$ a (real) Lie algebra; then $J^1_p\mathfrak{g}$ inherits a Lie algebra structure: for any $j^1(f), j^1(g) \in J^1_p\mathfrak{g}$ define

$$[j^1(f), j^1(g)] = j^1[f,g] ,$$

where $[f,g]: \mathbf{R}^p \longrightarrow \mathfrak{g}$ is given by $[f,g](t) = [f(t), g(t)]$, $t \in \mathbf{R}^p$.

In fact, if $\{\lambda^c_{ab}\}$ are the structure constants of $\mathfrak{g}$ with respect to $\{e_a\}$, i.e. $[e_a, e_b] = \lambda^c_{ab}e_c$, then the structure constants of $J^1_p\mathfrak{g}$ with respect to $\{E_a, E_{a_\alpha}\}$ are

$$\Lambda^a_{b.c.} = \lambda^a_{bc} \quad , \quad \Lambda^{c_\alpha}_{ab_\beta} = \delta^\alpha_\beta \lambda^c_{ab} ,$$
$$\Lambda^c_{ab_\alpha} = \Lambda^{c_\alpha}_{ab} = \Lambda^c_{a_\beta b_\gamma} = \Lambda^{c_\alpha}_{a_\beta b_\gamma} = 0 .$$

Let $Ad: \mathfrak{g} \longrightarrow Der(\times_p\mathfrak{g})$ be the canonical extension of the adjoint representation of $\mathfrak{g}$ to the Abelian Lie algebra $\times_p\mathfrak{g}$, and construct the semidirect product Lie algebra $\mathfrak{g}\times_{Ad}(\times_p\mathfrak{g})$; then

Lemma 1.3.1 *The isomorphism of vector spaces $J^1_p\mathfrak{g} \simeq \mathfrak{g}\times_{Ad}(\times_p\mathfrak{g})$ is an isomorphism of Lie algebras.*

Proof. Routine, from the definition of bracket products in $\mathfrak{g}\times_{Ad}(\times_p\mathfrak{g})$ and 1.2.□

Combining Theorem 1.2.1 and Lemma 1.3.1, it follows:

Corollary 1.3.2 *Let G be a Lie group, $\mathfrak{g}$ its Lie algebra. Then the Lie algebra of J^1_pG is canonically isomorphic to the Lie algebra $J^1_p\mathfrak{g}$.*

1.4 The embedding j_p

Let V be again an m–dimensional (real) vector space, as in Section 1.3, and denote by $Gl(V)$ the Lie group of automorphisms of V.

Let $\rho: Gl(V) \times V \longrightarrow V$ be the canonical action of $Gl(V)$; according to 1.2, $J^1_pGl(V)$ acts on the vector space J^1_pV through the induced map

$$\begin{aligned} \rho^1: J^1_pGl(V) \times J^1_pV &\longrightarrow J^1_pV \\ (j^1(f), j^1(\eta)) &\longmapsto j^1(f * \eta) \quad \text{(cf. 1.1)} \end{aligned}$$

where

$$f * \eta: \mathbf{R}^p \longrightarrow V \quad : \quad t \longmapsto f(t)(\eta(t)) .$$

Definition 1.4.1 *We introduce the map*

$$\begin{aligned} j_p\colon J_p^1 Gl(V) &\longrightarrow Gl(J_p^1 V) \\ j^1(f) &\longmapsto [j^1(\eta) \longmapsto j^1(f * \eta)] \,. \end{aligned}$$

The following theorem is proved by direct computations.

Theorem 1.4.2 *j_p is an embedding of Lie groups.*

With respect to the canonical coordinates of $J_p^1 Gl(V)$ and $Gl(J_p^1 V)$ the embedding j_p is given as follows: let $\{x_b^a\}$ be the global coordinate functions on $Gl(V)$ with respect to the basis $\{e_a\}$ of V, $\{x^a\}$ the coordinate functions on V with respect to the same basis, $\{x^a, x_\alpha^a\}$ the induced coordinates on $J_p^1 V$, $\{x_b^a, x_{b_\alpha}^a\}$ the induced coordinates on $J_p^1 Gl(V)$ and, finally, denote by $\{X_A^B, 1 \leq A, B \leq m{+}mp\}$ the natural coordinates on $Gl(J_p^1 V)$ with respect to the basis $\{E_a, E_{a_\alpha}\}$ induced on $J_p^1 V$. Using 1.1, a direct computation leads to the following equation:

$$\rho^1((x_b^a, x_{b_\alpha}^a), (x^c, x_\alpha^c)) = (x_c^a x^c, x_{c_\alpha}^a x^c + x_c^a x_\alpha^c) \,,$$

from where one easily obtains the following equations defining j_p:

$$(1.11) \qquad j_p\colon \left\{ \begin{array}{lllllll} X_b^a &=& x_b^a &,& X_{b_\alpha}^a &=& 0 \,, \\ X_b^{a_\alpha} &=& x_{b_\alpha}^a &,& X_{b_\beta}^{a_\alpha} &=& \delta_\beta^\alpha x_b^a \,. \end{array} \right.$$

$V = \mathbf{R}^n$

Let us consider $V = \mathbf{R}^n$; then $Gl(V) = Gl(n, \mathbf{R})$ and, taking into account (1.7), (1.9) and (1.10), one gets:

$$[a; B_\alpha] \cdot [x; X_\alpha] \mapsto \begin{pmatrix} a_i^h x^i \\ (B_\alpha)_j^h a_i^j x^i + a_i^h X_\alpha^i \end{pmatrix} \in \mathbf{R}^{n+pn},$$

$[x; X_\alpha] \in J_p^1 \mathbf{R}^n$, $[a; B_\alpha] \in J_p^1 Gl(n, \mathbf{R})$, $a \in Gl(n, \mathbf{R})$ and $B_\alpha \in gl(n, \mathbf{R})$. Therefore,

$$j_p\colon J_p^1 Gl(n, \mathbf{R}) \longrightarrow Gl(\mathbf{R}^{n+pn})$$

is given by

$$(1.12) \qquad j_p([a; B_\alpha]) = \begin{pmatrix} a & 0 & \dots & 0 \\ B_1 a & a & \dots & 0 \\ \vdots & \vdots & \ddots & \vdots \\ B_p a & 0 & \dots & a \end{pmatrix} .$$

Hence, as a consequence of Theorem 1.2.1 (or of Lemma 1.3.1), and of (1.12), there obtains:

Proposition 1.4.3 *Let G be a Lie subgroup of $Gl(n, \mathbf{R})$. Then the Lie algebra of $j_p(J_p^1 G)$ consists of all the matrices of the form*

$$\begin{pmatrix} A & 0 & \dots & 0 \\ B_1 & A & \dots & 0 \\ \vdots & \vdots & \ddots & \vdots \\ B_p & 0 & \dots & A \end{pmatrix}$$

for any $A, B_1, \dots, B_p \in \mathfrak{g}$, $\mathfrak{g}$=Lie algebra of G.

Remark 1.4.4 All the previous results are well known for $p = 1$ (see Morimoto [67]).

Chapter 2

Prolongation of G–structures to FM

Introduction

We have seen that FM is an open submanifold of J^1_nM, now we show that J^1_nFM embeds nicely in FFM. Next, a G–structure on M is a reduction to G of the structure group $Gl(n,\mathbf{R})$ of FM. However, there is a Lie group homomorphism of $J^1_nGl(n,\mathbf{R})$ into $Gl(n+n^2,\mathbf{R})$. Then we find that it preserves G–structures, so we can prolong G–structures on M to FM. Now, a G–structure is called integrable, or flat, if it contains a local canonical coordinate frame about each point of M. We see that integrability is preserved and co-preserved by our prolongation of G–structure. Endomorphisms of $\mathbf{R}^n$ and bilinear forms on $\mathbf{R}^n$ generate isotropy subgroups of $Gl(n,\mathbf{R})$ and hence G–structures, which we examine.

2.1 Imbedding of J^1_nFM into FFM

Let $P(M,\pi,G)$ be a principal fibre bundle with total space P, base space M, projection π and structure group G.

If $\{U\}$ is an open covering of M, P giving a trivial bundle over U, and if $g_{UU'}: U \cap U' \to G$ are the transition functions of P, we express this fibre bundle by $P = \{U, g_{UU'}\}$.

When G is a Lie subgroup of a Lie group G' and $j: G \to G'$ is the injection map, then there is a fibre bundle $P' = \{U, j \circ g_{UU'}\}$ and an injection $j: P \longrightarrow P'$ which is a bundle homomorphism, i.e. $j(p \cdot a) = j(p) \cdot a$ for any $p \in P$ and $a \in G$. If N is an open submanifold of M, then $P_{|N}$ will denote the restriction of P to N.

Proposition 2.1.1 *Let $P(M,\pi,G)$ be a principal fibre bundle, M an n–dimensional manifold. Then $J^1_nP(J^1_nM,\pi^1,J^1_nG)$ is again a principal fibre bundle, and* *if* $P = \{U, g_{UU'}\}$ *then* $J^1_n P = \{J^1_n U, g^1_{UU'}\}$.

Proof. Let $\phi_U : U \times G \to \pi^{-1}(U)$ be the trivialization of P over U. Then, by definition

$$\phi_U^{-1} \circ \phi_{U'}(x, g) = (x, g_{UU'}(x)g) \ , \ x \in U \cap U' \ , \ x \in G \ .$$

Since $(\pi^1)^{-1}(J^1_n U) = J^1_n(\pi^{-1}U)$, it suffices to prove the following:

$$(\phi^1_U)^{-1} \circ \phi^1_{U'}(x', g') = (x', g^1_{UU'}(x')g')$$

for $(x', g') \in (J^1_n U \cap J^1_n U') \times J^1_n G$. To prove this it is enough to prove the following assertion:

Let $f: U \to G$ *be a differentiable map and define*

$$\begin{aligned} \psi: U \times G &\longrightarrow U \times G \\ (x, g) &\longmapsto (x, f(x)g) \ . \end{aligned}$$

Then $\psi^1(x', g') = (x', f^1(x')g')$ *for* $(x', g') \in J^1_n U \times J^1_n G$.

Let $\mu: G \times G \to G$ be the group multiplication, and let

$$\pi_1: U \times G \to U \quad , \quad \pi_2: U \times G \to G$$

be the projections. Since

$$\pi_2 \circ \psi(x, g) = f(x)g = (\mu \circ (f \times 1_G))(x, g) \ ,$$

then $\mu_2 \circ \psi = \mu \circ (f \times 1_G)$ and hence

$$\begin{aligned} \pi^1_2 \circ \psi^1 &= (\pi_2 \circ \psi)^1 = (\mu \circ (f \times 1_G))^1 \\ &= \mu^1 \circ (f \times 1_G)^1 = \mu^1 \circ (f^1 \times 1_{J^1_n G}) \ . \end{aligned}$$

Therefore,

$$\pi^1_2 \circ \psi^1(x', g') = \mu^1(f^1(x'), g') = f^1(x')g' \ .$$

On the other hand,

$$\pi^1_1 \circ \psi^1(x', g') = (\pi_1 \circ \psi)^1(x', g') = \pi^1_1(x', g') = x' \ .$$

Thus, $\psi^1(x', g') = (x', f^1(x')g')$ and the proposition is proved.□

Let $FM(M, \pi_M, Gl(n, \mathbf{R}))$ denote the frame bundle of the differentiable manifold M, $J^1_n FM(J^1_n M, \pi^1_M, J^1_n Gl(n, \mathbf{R}))$ the induced $J^1_n Gl(n, \mathbf{R})$-principal bundle, and let $FJ^1_n M(J^1_n M, \pi_{J^1_n M}, Gl(n + n^2, \mathbf{R}))$ be the frame bundle of $J^1_n M$.

Theorem 2.1.2 *There exists a canonical injective homomorphism of principal bundles*

$$j_M: J_n^1FM \longrightarrow FJ_n^1M$$

over the identity of J_n^1M, with associated Lie group homomorphism

$$j_n: J_n^1Gl(n,\mathbf{R}) \longrightarrow Gl(n+n^2,\mathbf{R}) \quad .$$

Proof. Let U be a coordinate neighbourhood in M, and denote by

$$\phi_U: FU \times J_n^1Gl(n,\mathbf{R}) \longrightarrow J_n^1FU$$
$$\psi_U: J_n^1U \times Gl(n+n^2,\mathbf{R}) \longrightarrow FJ_n^1U$$

the local trivializations of J_n^1FM and FJ_n^1M respectively. Then we define a map

$$(j_M)_U: J_n^1FU \longrightarrow FJ_n^1U$$

as the composition

$$(j_M)_U = \psi_U \circ (i \times j_n) \circ \phi_U^{-1}$$

where $i: FU \to J_n^1U$ is the canonical injection. In order to prove that j_M is well defined from these local $(j_M)_U$ we only need to show that

$$(j_M)_{U|U\cap U'} = (j_M)_{U'|U\cap U'} \ , \tag{2.1}$$

for any coordinate neighbourhoods U, U' in M with $U \cap U' \neq \emptyset$. And to prove that (2.1) holds, it suffices to prove that one has

$$j_n \circ J'_{UU'} = \bar{J}_{UU'}$$

on $F(U \cap U') \subset J_n^1U \cap J_n^1U'$,where

$$J_{UU'} \ : \ U \cap U' \to Gl(n,\mathbf{R}) \ ,$$
$$\bar{J}_{UU'} \ : \ J_n^1U \cap J_n^1U' \to Gl(n+n^2,\mathbf{R})$$

denote the Jacobian matrices of change of coordinates in M and J_n^1M respectively.

To make the computation easier, let us denote by (x^i) the coordinates on U, (y^i) the coordinates on U', $J = J_{UU'}$, $\bar{J} = \bar{J}_{UU'}$, (x^i, x^i_α) the induced coordinates in J_n^1U and (X^i_j) the canonical coordinates on $Gl(n,\mathbf{R})$.

Then, if we put $y^i = f^i(x^1,\ldots,x^n)$ on $U \cap U'$,

$$J(x) = \left(\left(\frac{\partial f^i}{\partial x^i}\right)_x\right) \ .$$

Now, take an arbitrary $[x; X_\alpha] \in F(U \cap U')$ with $X_\alpha = x^i_\alpha (\partial/\partial x^i)_x$. Then $J^1([x; X_\alpha]) = [J(x); J_*(X_\alpha)]$ and, since

$$J_*(X_\alpha) = x^i_\alpha \frac{\partial^2 f^k}{\partial x^i \partial x^j}\left(\frac{\partial}{\partial X^k_j}\right)_{j(x)},$$

it follows

$$J^1([x; X_\alpha]) = \left[J(x); x^i_\alpha \frac{\partial^2 f^k}{\partial x^i \partial x^j}(J(x)^{-1})^j_h \left(\frac{\partial}{\partial X^k_h}\right)_e\right].$$

Therefore

$$j_n(J^1[x; X_\alpha]) = \begin{pmatrix} \left(\frac{\partial f^k}{\partial x^j}\right) & 0 & \cdots & 0 \\ \left(x^i_1 \frac{\partial^2 f^k}{\partial x^i \partial x^j}\right) & \left(\frac{\partial f^k}{\partial x^j}\right) & \cdots & 0 \\ \vdots & \vdots & \ddots & \vdots \\ \left(x^i_n \frac{\partial^2 f^k}{\partial x^i \partial x^j}\right) & 0 & \cdots & \left(\frac{\partial f^k}{\partial x^j}\right) \end{pmatrix}.$$

Taking into account (1.2), the theorem follows easily.□

If we now denote by $J^1_n FM_{|FM}$ the restriction of $J^1_n FM$ to the open submanifold FM of $J^1_n M$, and remark that the restriction $FJ^1_n M_{|FM}$ of $FJ^1_n M$ to the submanifold FM is canonically isomorphic, as a $Gl(n+n^2, \mathbf{R})$-bundle, to the frame bundle FFM of FM, then from Theorem 2.1.2 we deduce

Theorem 2.1.3 *Let be $\pi_M: FM \to M$ and $\pi_{FM}: FFM \to FM$ the frame bundles of M and FM, respectively. Then, j_M induces a bundle homomorphism of $J^1_n FM_{|FM}$ into FFM with respect to j_n, i.e.*

$$j_M(X \cdot Y) = j_M(X) \cdot j_n(Y)$$

for $X \in J^1_n FM_{|FM}$ and $Y \in J^1_n Gl(n, \mathbf{R})$, and the following diagram

$$\begin{array}{ccc} J^1_n FM_{|FM} & \xrightarrow{j_M} & FFM \\ \downarrow{\scriptstyle \pi^1_M} & & \downarrow{\scriptstyle \pi_{FM}} \\ FM & \xrightarrow{1_{FM}} & FM \end{array}$$

is commutative.

2.2 Prolongation of G–structures to FM

Let G be a Lie subgroup of $Gl(n, \mathbf{R})$. We shall denote by $\tilde{G}$ the image of $J^1_n G$ by j_n, that is $\tilde{G} = j_n(J^1_n G)$. Then, $\tilde{G}$ is a Lie subgroup of $Gl(n+n^2, \mathbf{R})$ isomorphic to $J^1_n G$.

Let us recall that a G–structure on an n–dimensional manifold M is a G–subbundle $P(M, \pi, G)$ of the frame bundle FM of M. Therefore, a G–structure on M is precisely a reduction of the structure group $Gl(n, \mathbf{R})$ of FM to the subgroup G.

Theorem 2.2.1 *If a manifold M possesses a G–structure P, then FM possesses a canonically induced $\tilde{G}$–structure $\tilde{P}$.*

Proof. Let us consider the $J^1_n G$–bundle $J^1_n P(J^1_n M, \pi^1, J^1_n G)$ and let $J^1_n P_{|FM}$ be its restriction to FM. Then, by virtue of Theorem 2.1.3, the image $\tilde{P} = j_M(J^1_n P_{|FM})$ is a subbundle of FFM with structure group $\tilde{G}$ and, hence, it defines a $\tilde{G}$–structure on FM.□

Definition 2.2.2 *We shall call $\tilde{P}$ in* Theorem 2.2.1 *the prolongation of the G–structure P on M to the frame bundle FM.*

Let M and M' be manifolds of dimension n, $f: M \longrightarrow M'$ a diffeomorphism of M onto M', $f^1: J^1_n M \longrightarrow J^1_n M'$ the diffeomorphism induced by f and $Ff = f^1{}_{|FM}: FM \longrightarrow FM'$ the diffeomorphism between the frame bundles induced by restriction of f^1 (or equivalently induced by f). Let G be a Lie subgroup of $Gl(n, \mathbf{R})$ and let P and P' be G-structures on M and M' respectively. The diffeomorphism $f: M \longrightarrow M'$ is said to be an isomorphism of P to P' if $(Ff)(P) = P'$.

Theorem 2.2.3 *Let $f: M \longrightarrow M'$ be a diffeomorphism. Then the following diagram commutes*

$$\begin{array}{ccc} J^1_n FM_{|FM} & \xrightarrow{\ j_M\ } & FJ^1_n M \\ {\scriptstyle (Ff)^1}\downarrow & & \downarrow{\scriptstyle Ff^1} \\ J^1_n FM' & \xrightarrow{\ j_{M'}\ } & FJ^1_n M' \end{array}$$

Proof. Let $\{U\}$ be an open covering of M by coordinate neighbourhoods U. Then $\{U' = f(U)\}$ is an open covering of M' by coordinate neighbourhoods U'. Let

$$\phi_U \ : \ U \times Gl(n, \mathbf{R}) \longrightarrow FU ,$$

$$\phi_{U'} \ : \ U' \times Gl(n, \mathbf{R}) \longrightarrow FU' ,$$

be the trivialization of FM and FM' over U and U' respectively. Define a diffeomorphism $f_{UU'}$ by setting

$$f_{UU'} = \phi_{U'}^{-1} \circ Ff \circ \phi_U : U \times Gl(n, \mathbf{R}) \longrightarrow U' \times Gl(n, \mathbf{R}) .$$

Then $f_{UU'}$ induces a diffeomorphism

$$f^1_{UU'} = (\phi_{U'}^{-1})^1 \circ (Ff)^1 \circ \phi^1_{U'} : J^1_n U \times J^1_n Gl(n, \mathbf{R}) \to J^1_n U' \times J^1_n Gl(n, \mathbf{R}) .$$

Now, let

$$\begin{aligned} \psi_U &: J^1_n U \times Gl(n+n^2, \mathbf{R}) \longrightarrow FJ^1_n U , \\ \psi_{U'} &: J^1_n U' \times Gl(n+n^2, \mathbf{R}) \longrightarrow FJ^1_n U' , \end{aligned}$$

be the local trivializations of $FJ^1_n M$ and $FJ^1_n M'$ over $J^1_n U$ and $J^1_n U'$, respectively, and define

$$\bar{f}_{UU'} : J^1_n U \times Gl(n+n^2, \mathbf{R}) \to J^1_n U' \times Gl(n+n^2, \mathbf{R})$$

by setting

$$\bar{f}_{UU'} = \psi_{UU'}^{-1} \circ Ff^1 \circ \psi_U \quad .$$

Finally, let

$$j_U = 1_{J^1_n U} \times j_n \quad , \quad j_{U'} = 1_{J^1_n U'} \times j_n$$

and recall that, by definition of j_M and $j_{M'}$,

$$\begin{aligned} j_M &= \psi_U \circ j_U \circ (\phi_U^{-1})^1 \quad \text{on} \quad J^1_n FU , \\ j_{M'} &= \psi_{U'} \circ j_{U'} \circ (\phi_{U'}^{-1})^1 \quad \text{on} \quad J^1_n FU' . \end{aligned}$$

Then, to prove the commutativity of the diagram we are reduced to proving that

$$j_{U'} \circ f^1_{UU'} = \bar{f}_{UU'} \circ j_{U'} \tag{2.2}$$

In order to prove this identity, we consider local coordinates (x^i) on U and (y^i) on U'. Then $J^1_n U$ (and $J^1_n U'$) has induced coordinates (x^i, x^i_α) (resp. (y^i, y^i_α)) and we shall denote by the same symbols the induced coordinates on $FU \subset J^1_n U$ (resp. $FU' \subset J^1_n U'$). If (X^i_j) are the global coordinates on $Gl(n, \mathbf{R})$, then $(X^i_j, X^i_{j\alpha})$ will denote the induced coordinates in $J^1_n Gl(n, \mathbf{R})$.

Let $f: U \to U'$ be expressed by $y^i = f^i(x^1, \ldots, x^n)$ in terms of the coordinate systems above. Then the maps $f_{UU'}$ and $f^1_{UU'}$ are locally expressed as follows:

$$f_{UU'}: \;\; y^i = f^i(x) \quad , \quad Y^i_j = \frac{\partial f^i}{\partial x^k} X^k_j \quad ; \tag{2.3}$$

$$(2.4)\qquad f^1_{UU'}: \begin{cases} y^i &= f^i(x)\,, \quad Y^i_j = \dfrac{\partial f^i}{\partial x^k} X^k_j\,, \\ y^i_\alpha &= \dfrac{\partial f^i}{\partial x^k} x^k_\alpha\,, \\ Y^i_{j_\alpha} &= X^k_j \dfrac{\partial^2 f^i}{\partial x^k \partial x^h} x^h_\alpha + \dfrac{\partial f^i}{\partial x^k} X^k_{j_\alpha}\,. \end{cases}$$

On the other hand, the coordinates on $J^1_n U$ and on $J^1_n U'$ induce coordinates

$$\{x^i, x^i_\alpha, \overline{X}^A_B\}\,,\ \{y^i, y^i_\alpha, \overline{Y}^A_B\}\,,\ A, B = 1, 2, \ldots, n + n^2\,.$$

Since

$$\overline{Y}^A_B = \overline{X}^C_B \frac{\partial f^A}{\partial x^C}\,,$$

if we put

$$f^{\alpha n+i}(x^j, x^j_\alpha) = \frac{\partial f^i}{\partial x^k} x^k_\alpha\,, \qquad x^{\alpha n+i} = x^i_\alpha\,,$$

then we can express $\bar{f}_{UU'}$ by the following equations:

$$(2.5)\qquad \bar{f}_{UU'}: \begin{cases} y^i = f^i(x)\,, \quad y^i_\alpha = \dfrac{\partial f^i}{\partial x^k} x^k_\alpha\,, \\ \overline{Y}^j_B = \overline{X}^k_B \dfrac{\partial f^j}{\partial x^k}\,, \\ \overline{Y}^{\alpha n+j}_B = \overline{X}^k_B \dfrac{\partial^2 f^j}{\partial x^k \partial x^h} x^h_\alpha + \overline{X}^{\alpha n+k}_B \dfrac{\partial f^j}{\partial x^k}\,. \end{cases}$$

We now assume that $\sigma \in J^1_n Gl(n, \mathbf{R})$ has coordinates $\{X^i_j, X^i_{j_\alpha}\}$; then, using (1.11), one easily gets

$$(2.6)\qquad j_n(\sigma) = \begin{pmatrix} (X^i_j) & 0 & \ldots & 0 \\ (X^i_{j_1}) & (X^i_j) & \ldots & 0 \\ \vdots & \vdots & \ddots & \vdots \\ (X^i_{j_n}) & 0 & \ldots & (X^i_j) \end{pmatrix}.$$

Now, let be $\tilde{p} = (x^i, x^i_\alpha, X^i_j, X^i_{j_\alpha}) \in J^1 U \times J^1_n Gl(n, \mathbf{R})$; then, in view of (2.3) and (2.4)

$$f^1_{UU'}(\tilde{p}) = (f^i(x), \frac{\partial f^i}{\partial x^k} x^k_\alpha, Y^i_j, Y^i_{j_\alpha})\,,$$

where Y^i_j, $Y^i_{j_\alpha}$ are given by (2.4).

Next, using (2.5) and (2.6), and by direct computation, we obtain (2.2) and the theorem is proved.□

Theorem 2.2.4 *Let P and P' be G-structures on M and M', respectively, and let $f: M \to M'$ be a diffeomorphism of M onto M'. Then f is an isomorphism of P to P' if and only if Ff is an isomorphism of $\tilde{P}$ to $\tilde{P}'$.*

Proof. By definition $\tilde{P} = j_M(J^1_n P_{|FM})$ and $\tilde{P}' = j_{M'}(J^1_n P'_{|FM'})$. Suppose that f is an isomorphism of P to P'. Then

$$\begin{aligned} FFf(\tilde{P}) &= FFf \circ j_M(J^1_n P_{|FM}) \\ &= j_{M'}((Ff)^1(J^1_n P_{|FM})) \\ &= j_{M'}(J^1_n P'_{|FM'}) \\ &= \tilde{P}' , \end{aligned}$$

since $FFf = Ff^1{}_{|FM}$ and

$$(Ff)^1(J^1_n P_{|FM}) = J^1_n P'_{|FM'} .$$

Therefore Ff is an isomorphism of $\tilde{P}$ to $\tilde{P}'$.

Conversely, suppose that Ff is an isomorphism of $\tilde{P}$ to $\tilde{P}'$. Then

$$FFf(\tilde{P}) = \tilde{P}' \quad , \quad FFf(j_M(J^1_n P_{|FM})) = j_{M'}(J^1_n P'_{|FM'}) .$$

Therefore, $j_{M'}((Ff)^1(J^1_n P_{|FM})) = j_{M'}(J^1_n P_{|FM'})$. Since $j_{M'}$ is injective,

$$(Ff)^1(J^1_n P_{|FM}) = J^1_n P'_{|FM'}$$

which implies $Ff(P) = P'$, whence f is an isomorphism of P onto P'.□

Corollary 2.2.5 *Let P be a G-structure on M and let f be a diffeomorphism of M onto itself. Then, f is an automorphism of P if and only if Ff is an automorphism of $\tilde{P}$.*

On the other hand, let us recall that given a G-structure P on M, a vector field X on M is called an infinitesimal automorphism of P if it generates a local 1-parameter group of local transformations of P. Then, using the results in 1.1, the following corollary is of easy proof.

Corollary 2.2.6 *Let P be a G-structure on M, and let $\tilde{P}$ be its prolongation to FM. Then a vector field X on M is an infinitesimal automorphism of P if and only if its complete lift X^C to FM is an infinitesimal automorphism of $\tilde{P}$.*

2.3 Integrability

Let us begin by recalling the following definition.

Definition 2.3.1 *Let $P(M,\pi,G)$ be a G-structure on M. P is said to be integrable (or flat) if for each point $x \in M$ there is a chart (U, x^i) with $x \in U$ such that the canonical frame*

$$\left(\left(\frac{\partial}{\partial x^1}\right)_y, \dots, \left(\frac{\partial}{\partial x^n}\right)_y\right) \in P$$

for every $y \in U$.

Lemma 2.3.2 *Let $(x^1, \dots, x^n)$ be coordinates on a neighbourhood U in M, and let $f: U \longrightarrow Gl(n, \mathbf{R})$ be a map, $f^i_j(x)$ being the (i,j)-entry of $f(x)$ for $x \in U$. Then, if $[x; X_\alpha] \in J^1_n U$ has coordinates (x^i, x^i_α),*

$$(2.7) \qquad (j_n \circ f^1)([x; X_\alpha]) = \begin{pmatrix} (f^i_j(x)) & 0 & \dots & 0 \\ \left(\frac{\partial f^i_j}{\partial x^k} x^k_1\right) & (f^i_j(x)) & \dots & 0 \\ \vdots & \vdots & \ddots & \vdots \\ \left(\frac{\partial f^i_j}{\partial x^k} x^k_n\right) & 0 & \dots & (f^i_j(x)) \end{pmatrix}.$$

Proof. Let X^i_j denote again the natural coordinate functions of $Gl(n, \mathbf{R})$. Then

$$f^1([x; X_\alpha]) = [(f^i_j(x)); f_* X_\alpha] = \left[(f^i_j(x)); x^k_\alpha \frac{\partial f^i_j}{\partial x^k} \left(\frac{\partial}{\partial X^i_j}\right)_{f(x)}\right].$$

Now, applying (2.6), (2.7) follows directly.□

Proposition 2.3.3 *Let $(x^1, \dots, x^n)$ be local coordinates on $U \subset M$. Let*

$$\begin{aligned} \phi: U &\longrightarrow FM \\ x &\longmapsto \left(\phi^i_j(x) \left(\frac{\partial}{\partial x^i}\right)_x\right) \end{aligned}$$

be a local section. Define

$$\tilde{\phi} = j_M \circ \phi^1 .$$

Then $\tilde{\phi}$ is a section of $FJ^1_n M$ over $J^1_n U$ and is expressed by

$$\tilde{\phi}(X) = \left(\phi^i_j(x) \left(\frac{\partial}{\partial x^i}\right)_X + \frac{\partial \phi^i_j}{\partial x^k} x^k_\alpha \left(\frac{\partial}{\partial x^i}\right)_X, \ \phi^i_j(x) \left(\frac{\partial}{\partial x^i_\alpha}\right)_X\right)$$

for any $X = [x; X_\alpha] \in J^1_n U$, $x = (x^1, \ldots, x^n)$, $X_\alpha = x^i_\alpha(\partial/\partial x^i)_x$, *and* (x^i, x^i_α) *being the induced coordinates on* $J^1_n U$. *In particular,*

$$\bar{\phi} = \tilde{\phi}_{|FU}$$

is a section of FFM over FU.

Proof. Let $\pi_M: FM \to M$ and $\pi_{J^1_n M}: FJ^1_n M \to J^1_n M$ be the projections, and let ϕ_U and ψ_U be the local trivializations of FM and $FJ^1_n M$ over U and $J^1_n U$ respectively. Then, by Theorem 2.2.3,

$$j_{M|J^1_n FU} = \psi_U \circ (i \times j_n) \circ (\phi^1_U)^{-1}$$

where $i: FU \longrightarrow J^1_n U$ is the canonical injection. Firstly,

$$\pi_{J^1_n M} \circ \tilde{\phi} = \pi_{J^1_n M} \circ j_M \circ \phi^1 = \pi^1_M \circ \phi^1 = (\pi_M \circ \phi)^1 = 1_{J^1_n U} ,$$

which shows that ϕ is a section of $FJ^1_n M$ over $J^1_n U$.

Next, putting $f(x) = (\phi^i_j(x))$ for $x \in U$ and using Lemma 2.3.2, it follows

$$\begin{aligned} \tilde{\phi} &= \psi_U \circ (i \times j_n) \circ (\phi^1_U)^{-1} \circ (\phi_U \circ \phi_U^{-1} \circ \phi)^1 \\ &= \psi_U \circ (i \times j_n) \circ (1_U \times f)^1 \\ &= \psi_U \circ (i \times (j_n \circ f^1)) , \end{aligned}$$

which implies

$$\tilde{\phi}(X) = (x^i, x^i_\alpha, \widetilde{X}_A) ,$$

where

$$\begin{aligned} \widetilde{X}_j &= \phi^i_j(x) \left(\frac{\partial}{\partial x^i}\right)_X + \frac{\partial \phi^i_j}{\partial x^k} x^k_\alpha \left(\frac{\partial}{\partial x^i}\right)_X , \\ \widetilde{X}_{\alpha n+j} &= \phi^i_j(x) \left(\frac{\partial}{\partial x^i_\alpha}\right)_X , \end{aligned}$$

and the proposition is proved.□

Remark 2.3.4 Note that, in view of (1.3) and (1.4), the section $\tilde{\phi}$ in Proposition 2.3.3 is given as

$$\tilde{\phi}(X) = \left(X^C_j(X), X^{(\alpha)}_j(X)\right) , \quad X \in J^1_n U ,$$

where $\{X_j\}$ is the local field of frames defined by ϕ on U, i.e.

$$X_j = \phi^i_j \frac{\partial}{\partial x^i} .$$

Corollary 2.3.5 *Let P be an integrable G-structure on M; then the prolongation $\widetilde{P}$ of P to FM is also integrable.*

Proof. Take arbitrary $X_0 \in FM$, (U, x^i) a chart with $\pi_M(X_0) \in U$ such that the canonical section $\phi(x) = (\partial/\partial x^i)_x$ takes values in P for every $x \in U$.

Then in view of Proposition 2.3.3 and Remark 2.3.4, we deduce that $\bar{\phi} = \tilde{\phi}_{|FU}$ takes values in $\widetilde{P}$; but $\bar{\phi}$ is the canonical section associated to the induced chart (FU, x^i, x^i_α), hence $\widetilde{P}$ is integrable.□

Conversely,

Proposition 2.3.6 *Let P be a G-structure on M. If the prolongation $\widetilde{P}$ of P is integrable, then P is integrable.*

Proof. Let $x_0 \in M$ be an arbitrary point, (U, x^i) the local chart around x_0, and $\phi: U \longrightarrow P$ a local section. Then, by Proposition 2.3.3, $\bar{\phi} = j_M \circ \phi^1{}_{|FU}$ is a local section of $\widetilde{P}$ over FU.

Now, let $X_0 \in FU$ be the linear frame at x_0 defined by

$$X_0 = \left(\left(\frac{\partial}{\partial x^1} \right)_{x_0}, \ldots, \left(\frac{\partial}{\partial x^n} \right)_{x_0} \right) ,$$

Since $\widetilde{P}$ is integrable, there can be found a coordinate neighbourhood $\widetilde{U}$ around X_0 with coordinates $(y^A) = (y^1, \ldots, y^{n+n^2})$ such that $\widetilde{U} \subset FU$ and that, if we define $\tilde{\phi}_0$ by

$$\tilde{\phi}_0 = \left(\left(\frac{\partial}{\partial y^1} \right)_X, \ldots, \left(\frac{\partial}{\partial y^{n+n^2}} \right)_X \right) , \; X \in \widetilde{U} ,$$

then $\tilde{\phi}_0$ is a section of $\widetilde{P}$ over $\widetilde{U}$. Since $\bar{\phi}_{|\widetilde{U}}$ and $\tilde{\phi}_0$ both are sections of $\widetilde{P}$ over $\widetilde{U}$, there exists a map

$$\tilde{g}: \widetilde{U} \longrightarrow \widetilde{G} = j_n(J^1_n G)$$

such that

$$\tilde{\phi}_0(X) = \bar{\phi}(X) \cdot \tilde{g}(X) \, , \; X \in \widetilde{U} \, . \tag{2.8}$$

In view of 1.4 and (1.12), we can write

$$\tilde{g}(X) = \begin{pmatrix} g(X) & 0 & \ldots & 0 \\ B_1(X)g(X) & g(X) & \ldots & 0 \\ \vdots & \vdots & \ddots & \vdots \\ B_n(X)g(X) & 0 & \ldots & g(X) \end{pmatrix}$$

where $g: \widetilde{U} \longrightarrow G$ and $B_\alpha: \widetilde{U} \longrightarrow G$ are differentiable maps. Now, if we put

$$\begin{aligned}
\phi(x) &= \left(\phi^i_j(x)\left(\frac{\partial}{\partial x^i}\right)_x\right), \quad x \in U ,\\
g(X) &= \left(g^i_j(X)\right) ,\\
B_\alpha(X) &= \left(B^i_{j_\alpha}(X)\right) ,\\
g^i_{j_\alpha}(X) &= B^i_{k_\alpha}(X) g^k_j(X) , \quad X \in \widetilde{U} ,
\end{aligned}$$

then, using Remark 2.3.4, (2.8) can be written, for $X \in \widetilde{U}$, as follows: for each $X \in \widetilde{U}$

$$\begin{aligned}
\left(\frac{\partial}{\partial y^i}\right)_X &= (g^k_j X^C_k + g^i_{j_\alpha} X^{(\alpha)}_i)(X) ,\\
\left(\frac{\partial}{\partial y^{\alpha n+j}}\right)_X &= (g^k_j X^{(\alpha)}_k)(X) ,
\end{aligned}$$

from where we put

$$0 = \left[\frac{\partial}{\partial y^i}, \frac{\partial}{\partial y^{\alpha n+j}}\right] = -g^h_j X^{(\alpha)}_h (g^k_i) X^C_k + \textit{terms in } X^{(\alpha)}_j ,$$

and, since X^C_j, $X^{(\alpha)}_j$ are linearly independent in $\widetilde{U}$,

$$g^h_j X^{(\alpha)}_h (g^k_i) = 0 ;$$

therefore

$$X^{(\alpha)}_h (g^k_i) = 0 , \tag{2.9}$$

Since the matrix (ϕ^i_j) is non-singular at any $x \in U$, from (1.4) and (2.9) we deduce that $g = (g^i_j)$ does not depend on the coordinates x^i_α on $\widetilde{U}$. Hence there exists a family of functions α^i_j on $\pi_M(\widetilde{U}) \subset U$ such that

$$g^i_j = (\alpha^i_j \circ \pi_M)_{|\widetilde{U}} .$$

Now, define n vector fields on $\pi_M(\widetilde{U})$ by

$$W_j = \alpha^i_j X_i . \tag{2.10}$$

Then

$$[W_j, W_i] = \left(g^k_j \phi^l_k \frac{\partial g^h_i}{\partial x^l} - g^k_i \phi^l_k \frac{\partial g^h_j}{\partial x^l}\right) X_h .$$

On the other hand,

$$0 = \left[\frac{\partial}{\partial y^i}, \frac{\partial}{\partial y^j}\right] = \left(g^k_j \phi^l_k \frac{\partial g^h_i}{\partial x^l} - g^k_i \phi^l_k \frac{\partial g^h_j}{\partial x^l}\right) X^C_h + \textit{terms in } X^{(\alpha)}_j ,$$

and, therefore,

$$g_j^k \phi_k^l \frac{\partial g_i^h}{\partial x^l} - g_i^k \phi_k^l \frac{\partial g_j^h}{\partial x^l} = 0 \, .$$

Thus $[W_j, W_i] = 0$ and hence $\{W_1, \ldots, W_n\}$ is a natural frame on $\pi_M(\widetilde{U})$; that is, there exist coordinates $(\bar{x}^1, \ldots, \bar{x}^n)$ on $\pi_M(\widetilde{U})$ such that $W_i = \partial/\partial \bar{x}^i$. Now, from (2.9) and (2.10), it follows that

$$\left(\frac{\partial}{\partial \bar{x}^i}\right) = \alpha_j^i \phi_i^k \frac{\partial}{\partial x^k}$$

on $\pi_M(\widetilde{U})$, and therefore the section

$$\widehat{\phi}: \pi_M(\widetilde{U}) \longrightarrow FM$$

given by

$$\widehat{\phi}(x) = \left(\left(\frac{\partial}{\partial \bar{x}^1}\right)_x, \ldots, \left(\frac{\partial}{\partial \bar{x}^n}\right)_x\right) \, , \; x \in \pi_M(\widetilde{U}) \, ,$$

satisfies $\widehat{\phi}(x) = \bar{\phi}(x) g(x)$, where $g(x) = g(X)$ for any $X \in \widetilde{U}$ with $\pi_M(X) = x$. Since $g(X) \in G$ for any $X \in \widetilde{U}$, it follows that $\widehat{\phi}$ is, in fact, a section of P over U and the proposition is proved.□

Combining Proposition 2.3.6 and Corollary 2.3.5, it follows

Theorem 2.3.7 *Let P be a G-structure on a manifold M. Then P is integrable if and only if its prolongation $\widetilde{P}$ to FM is integrable.*

2.4 Applications

Let P be a G-structure on M, (U, x^i) a chart on M, and $\phi: U \to P$ a local section of P over U. Then, ϕ defines a local field of frames adapted to P given by $\{X_j = \phi_j^i (\partial/\partial x^i)\}$; therefore, the local field of coframes dual to $\{X_j\}$ is given by

$$\theta^j = \psi_i^j dx^i \, , \tag{2.11}$$

where (ψ_i^j) denotes the inverse of the matrix (ϕ_i^j).

On the other hand, according to Remark 2.3.4, ϕ induces a local section

$$\bar{\phi}: FU \longrightarrow \widetilde{P}$$

given by

$$\bar{\phi}(X) = \{(X_j^C)_X, (X_j^{(\alpha)})_X\} \, , \; X \in FU \, .$$

Therefore, the local field of frames given by $\bar{\phi}$ and adapted to $\widetilde{P}$ is $\{X_j^C, X_j^{(\alpha)}\}$ and, by consequence, its corresponding field of dual coframes $\{\tilde{\theta}^A\}$ is given by

$$\tilde{\theta}^h = \psi_i^j \, dx^i \, , \quad \tilde{\theta}^{\alpha n+h} = \tilde{\theta}_\alpha^h = \frac{\partial \psi_i^h}{\partial x^k} x_\alpha^k \, dx^i + \psi_i^h \, dx_\alpha^i \, . \tag{2.12}$$

Linear endomorphisms

Let $\rho: Gl(n, \mathbf{R}) \longrightarrow Gl(\mathbf{R}^n)$ be the canonical representation, and take an arbitrary linear map $u: \mathbf{R}^n \to \mathbf{R}^n$, i.e. $u \in End(\mathbf{R}^n)$. Let G_u denote the isotropy group of u with respect to ρ, that is

$$G_u = \{a \in Gl(n, \mathbf{R}) \mid a \cdot u = u \cdot a\} .$$

If we now consider the induced linear map

$$\tilde{u} = j^1 u: J^1_n \mathbf{R}^n \longrightarrow J^1_n \mathbf{R}^n ,$$

where $J^1_n \mathbf{R}^n \equiv \mathbf{R}^{n+n^2}$, then we have

Lemma 2.4.1 *Let $\tilde{u} = j^1 u \in End\ (\mathbf{R}^{n+n^2})$ be the linear map induced by $u \in End(\mathbf{R}^n)$. If rank $u = r$, then rank $\tilde{u} = (n+1)r$. Moreover, if u satisfies the polynomial equation $Q(u) = 0$, then $\tilde{u}$ satisfies the same equation $Q(\tilde{u}) = 0$.*

Proof. Let $u = (u^i_j)$ be the matrix expression of u with respect to the canonical basis of $\mathbf{R}^n$. Then, the matrix expression of $\tilde{u}$ with respect to the canonical basis of $\mathbf{R}^{n+n^2}$ is easily checked to be

$$\tilde{u} = \begin{pmatrix} (u^i_j) & \dots & 0 \\ \vdots & \ddots & \vdots \\ 0 & \dots & (u^i_j) \end{pmatrix} \tag{2.13}$$

and the result follows.□

Proposition 2.4.2 *Let $G_{\tilde{u}}$ be the isotropy group of $\tilde{u} \in End(\mathbf{R}^{n+n^2})$ with respect to the canonical representation*

$$\rho: Gl(n+n^2, \mathbf{R}) \longrightarrow Gl(\mathbf{R}^{n+n^2})$$

and denote $\overline{G_u} = j_n(J^1_n G)$. Then $\overline{G_u} \subset G_{\tilde{u}}$.

Proof. From (1.12) we know that $\tilde{a} \in \overline{G_u}$ is of the form

$$\tilde{a} = \begin{pmatrix} a & 0 & \dots & 0 \\ B_1 a & a & \dots & 0 \\ \vdots & \vdots & \ddots & \vdots \\ B_n a & 0 & \dots & a \end{pmatrix}$$

for some $a \in G_u$ and $B_1, \dots, B_n \in \mathfrak{g}_u$, the Lie algebra of G_u. Then, it suffices to check that $\tilde{a} \cdot \tilde{u} = \tilde{u} \cdot \tilde{a}$ for any $\tilde{a} \in \overline{G_u}$.□

Corollary 2.4.3 *If M possesses a G_u-structure, then FM possesses a $G_{\tilde{u}}$-structure. Moreover, if the G_u-structure on M is integrable, the induced $G_{\tilde{u}}$-structure on FM is also integrable.*

Proof. From Theorem 2.2.1, if P is a G_u-structure on M then $\tilde{P}$ is a $G_{\tilde{u}}$-structure on FM which defines a $G_{\tilde{u}}$-structure $\tilde{P}'$ on FM through the canonical injection $\overline{G_u} \hookrightarrow G_{\tilde{u}}$. If P is integrable then, by Theorem 2.3.7, $\tilde{P}$ is integrable and hence $\tilde{P}'$ is also integrable.□

Let P be a G_u-structure on M; then, associated to P, there exists a tensor field F on M of type $(1,1)$ defined as follows: for any $x \in M$, let

$$z_x : \mathbf{R}^n \longrightarrow T_xM$$

be a linear frame at x such that $z_x \in P$; then

$$F_x : T_xM \longrightarrow T_xM$$

is given by

$$F_x = z_x \circ u \circ z_x^{-1} \, .$$

In fact, let (U, x^i) be a chart on M, $\phi : U \longrightarrow P$ a section of P over U given by

$$\phi(x) = \left(\phi^i_j(x) \left(\frac{\partial}{\partial x^i} \right)_x \right) ,$$

and $\{X_j\}$, $\{\theta^j\}$ the local field of frames adapted to P, and determined by ϕ; then F is given on U by

$$F = F^i_j \frac{\partial}{\partial x^i} \otimes dx^j \, ,$$

with respect to the canonical frames, or equivalently, by

$$F = u^i_j X_i \otimes \theta^j$$

with respect to the adapted local frames.

Similarly, associated to the $G_{\tilde{u}}$-structure $\tilde{P}'$ on FM, as obtained in Corollary 2.4.3 from the prolongation $\tilde{P}$ of P, there exists a tensor field $\tilde{F}$ on FM of type $(1,1)$ which is locally given in FU, with respect to the adapted fields of frames $\{X^C_j, X^{(\alpha)}_j\}$ and $\{\tilde{\theta}^h, \tilde{\theta}^h_\alpha\}$, by

$$\tilde{F} = u^h_i \, X^C_h \otimes \tilde{\theta}^i + \delta^\beta_\alpha u^j_i \, X^{(\alpha)}_j \otimes \tilde{\theta}^i_\beta \, ,$$

by virtue of (2.13).

Therefore, with respect to the canonical frames defined in the induced chart (FU, x^i, x^i_α), $\widetilde{F}$ is given by

$$
(2.14)\qquad \left\{
\begin{aligned}
\widetilde{F} &= \phi^i_h u^h_k \psi^k_j \frac{\partial}{\partial x^i} \otimes dx^j \\
&\quad + x^k_\alpha u^h_i \left(\frac{\partial \phi^j_h}{\partial x^k} \psi^i_l + \phi^j_h \frac{\partial \psi^i_l}{\partial x^k} \right) \frac{\partial}{\partial x^j_\alpha} \otimes dx^l \\
&\quad + \delta^\beta_\alpha \phi^j_h u^h_i \psi^i_l \frac{\partial}{\partial x^j_\alpha} \otimes dx^l_\beta \quad .
\end{aligned}
\right.
$$

Next, taking into account that $F^i_j = \phi^i_h u^h_l \psi^l_j$, we get

$$
\frac{\partial F^i_j}{\partial x^k} = u^h_l \left(\frac{\partial \phi^i_h}{\partial x^k} \psi^l_j + \phi^i_h \frac{\partial \psi^l_j}{\partial x^k} \right) ,
$$

and then, combining with (2.14), we finally obtain, by straightforward computation, the following local expression for $\widetilde{F}$ with respect to the coordinates (x^i, x^i_α) on FU:

$$
(2.15)\qquad \widetilde{F} = F^i_j \frac{\partial}{\partial x^i} \otimes dx^j + x^k_\alpha \frac{\partial F^i_j}{\partial x^k} \frac{\partial}{\partial x^i_\alpha} \otimes dx^j + \delta^\beta_\alpha F^i_j \frac{\partial}{\partial x^i_\alpha} \otimes dx^j_\beta \, .
$$

Comparing (2.15) with the so called *complete lift* F^C *of* F *to* FM due to Mok [65,p.78], one easily sees that $\widetilde{F} = F^C$.

Summing up, we can state the following

Theorem 2.4.4 *Let P be a G_u–structure on M, $u \in \mathrm{End}(\mathbf{R}^n)$, and let F be the tensor field on M of type $(1,1)$ defined from P. Then the tensor field $\widetilde{F}$ on FM of type $(1,1)$ defined from the $G_{\tilde{u}}$–structure induced by $\widetilde{P}$ on FM (by virtue of* Corollary 2.4.3*) is just the complete lift F^C of F to FM, that is $\widetilde{F} = F^C$.*

Remark 2.4.5 The theorem above justifies the naturality of the definition of F^C as given by Mok in [65]. A justification of the corresponding definition for the complete lift of tensor fields of type $(1, s)$, $s \geq 2$, will be obtained in Chapter 3, although it is obvious that the study above can be extended to those tensor fields in a natural way from the considerations explained in Section 3.2.

Corollary 2.4.6 [65] *Let F be a tensor field on M of type $(1,1)$ defining a polynomial structure of rank k. Then the complete lift F^C of F to FM defines a polynomial structure on FM with the same structural polynomial and of rank $(n+1)r$.*

Bilinear forms

Let $u: \mathbf{R}^n \times \mathbf{R}^n \longrightarrow \mathbf{R}$ be a bilinear map, that is $u \in \otimes_2(\mathbf{R}^n)^*$, and denote by G_u the isotropy group of u with respect to the canonical representation $\rho: Gl(n, \mathbf{R}) \longrightarrow Gl(\mathbf{R}^n)$, that is

$$G_u = \{a \in Gl(n, \mathbf{R}) \mid u(ax, ay) = u(x, y), x, y \in \mathbf{R}^n\} .$$

Let $j^1u: J^1_n(\mathbf{R}^n \times \mathbf{R}^n) \longrightarrow J^1_n\mathbf{R}$ be the map induced by u, and define a map

$$\pi: J^1_n\mathbf{R} \longrightarrow \mathbf{R}$$

by setting

$$\pi(s, c_1, \ldots, c_n) = \sum_{\alpha=1}^{n} c_\alpha ,$$

where $J^1_n\mathbf{R} \equiv \mathbf{R}^{1+n}$ in the obvious way. Then, taking into account the canonical identifications

$$J^1_n(\mathbf{R}^n \times \mathbf{R}^n) \equiv J^1_n\mathbf{R}^n \times J^1_n\mathbf{R}^n \equiv \mathbf{R}^{n+n^2} \times \mathbf{R}^{n+n^2} ,$$

we can define a map

$$\tilde{u}: \mathbf{R}^{n+n^2} \times \mathbf{R}^{n+n^2} \longrightarrow \mathbf{R}$$

as the composition $\tilde{u} = \pi \circ j^1u$.

Lemma 2.4.7 *$\tilde{u}$ is a bilinear map, i.e. $\tilde{u} \in \otimes_2(\mathbf{R}^{n+n^2})$. Moreover, if u is symmetric (resp. skew-symmetric) then $\tilde{u}$ is also symmetric (resp. skew-symmetric), and if* rank $u = r$, *then* rank $\tilde{u} = 2r$.

Proof. Take arbitrary points $x = (x^1, \ldots, x^n)$, $y = (y^1, \ldots, y^n) \in \mathbf{R}^n$ and $[x; X_\alpha]$, $[y; X_\alpha] \in J^1_n\mathbf{R}^n$, the latter having coördinates (x^i, x^i_α), (y^i, y^i_α), respectively.

If we put $u(x^i, y^i) = u_{ij}x^iy^j$, that is if $u = (u_{ij})$ is the matrix representation of u with respect to the canonical basis of $\mathbf{R}^n$, then

$$j^1u((x^i, x^i_\alpha), (y^i, y^i_\alpha)) = (u_{ij}x^iy^j, x^i_\alpha u_{ij}y^j + y^j_\alpha u_{ij}x^i) ;$$

therefore

$$\tilde{u}((x^i, x^i_\alpha), (y^i, y^i_\alpha)) = \sum_{\alpha=1}^{n}(x^i_\alpha u_{ij}y^j + y^j_\alpha u_{ij}x^i) .$$

Thus, the matrix expression of $\tilde{u}$ with respect to the canonical basis of $\mathbf{R}^{n+n^2}$ is

$$\tilde{u} = \begin{pmatrix} 0 & (u_{ij}) & \ldots & (u_{ij}) \\ (u_{ij}) & 0 & \ldots & 0 \\ \vdots & \vdots & \ddots & \vdots \\ (u_{ij}) & 0 & \ldots & 0 \end{pmatrix} \tag{2.16}$$

and the lemma follows easily. Note that, with the identification $J^1_n\mathbf{R}^n \equiv \mathbf{R}^{n+n^2}$ given by (1.10), we can write

$$\tilde{u}(X,Y) = X^t \cdot \tilde{u} \cdot Y,$$

where

$$X = \begin{pmatrix} x^i \\ x^i_\alpha \end{pmatrix} \quad , \quad Y = \begin{pmatrix} y^i \\ y^i_\alpha \end{pmatrix} . \ \square$$

Proposition 2.4.8 *Let $G_{\tilde{u}}$ be the isotropy group of $\tilde{u} = \pi \circ j^1 u$ with respect to the natural representation of $Gl(n+n^2, \mathbf{R})$ in $Gl(\mathbf{R}^{n+n^2})$, and denote $\overline{G_u} = j_n(J^1_n G_u)$. Then $\overline{G_u} \subset G_{\tilde{u}}$.*

Proof. Let $\mathfrak{g}_u$ be the Lie algebra of G_u; then the identity $Bu + uB^t = 0$ is satisfied for any $B \in \mathfrak{g}_u$. Let $\tilde{a} \in \overline{G_u}$; then, $\tilde{a}$ is of the form

$$\tilde{a} = \begin{pmatrix} a & 0 & \dots & 0 \\ B_1 a & a & \dots & 0 \\ \vdots & \vdots & \ddots & \vdots \\ B_n a & 0 & \dots & a \end{pmatrix}$$

for some $a \in G_u$ and $B_1, \dots, B_n \in \mathfrak{g}_u$. Thus, having in mind that $a^t u a = a$ for every $a \in G_u$, a straightforward computation proves the proposition.$\square$

Corollary 2.4.9 *If M possesses a G_u-structure then FM possesses an induced $G_{\tilde{u}}$-structure. Moreover, if the G_u-structure on M is integrable, then the induced $G_{\tilde{u}}$-structure on FM is also integrable.*

Proof. Similar to that of Corollary 2.4.3.$\square$

Recall that an $Sp(n, \mathbf{R})$-structure (ussually colled almost symplectic structure) on M is uniquely determined by a differential 2-form of maximal rank; moreover, the structure is symplectic or integrable if and only if the almost symplectic form is closed.

Corollary 2.4.10 *If M has an almost symplectic structure, then FM has an induced almost presymplectic structure of rank $2n$. Moreover, if M has a symplectic structure, then FM has a presymplectic structure, that is the induced 2-form on FM is also closed.*

Let $u \in \otimes_2(\mathbf{R}^n)^*$ and P a G_u-structure on M. Then, there exists a tensor field G on M of type $(0,2)$ induced by P and defined as follows: for any $x \in M$, let $z_x \colon \mathbf{R}^n \longrightarrow T_x M$ be a linear frame at x such that $z_x \in P$; now put

$$\begin{aligned} G_x \colon T_x M \times T_x M &\longrightarrow \mathbf{R} \\ (X, Y) &\longmapsto u(z_x^{-1}(X), z_x^{-1}(Y)) \ . \end{aligned}$$

In fact, if (U, x^i) is a chart on M, $\phi: U \longrightarrow P$ a section of P over U, then G is given on U by

$$G = G_{ij}\, dx^i \otimes dx^j ,$$

with respect to the canonical frames of fields or, equivalently, by

$$G = u_{ij}\, \theta^i \otimes \theta^j ,$$

with respect to the adapted local frames.

Similarly, associated to the $G_{\tilde{u}}$–structure $\widetilde{P}'$ induced by P on FM there exists a tensor field $\widetilde{G}$ on FM of type $(0,2)$ which is locally given in FU by

$$\widetilde{G} = \tilde{u}_{AB}\, \tilde{\theta}^A \otimes \tilde{\theta}^B ,$$

with respect to the local field of frames given in (2.12), $\tilde{u} = (\tilde{u}_{AB})$ being given by (2.16); that is

$$\widetilde{G} = \sum_{\alpha=1}^{n} \left(u_{ij}\, \tilde{\theta}^i \otimes \tilde{\theta}^j + u_{ij}\, \tilde{\theta}^i_\alpha \otimes \tilde{\theta}^j_\alpha \right) .$$

Therefore, with respect to the canonical coframes defined by the induced coordinates (x^i, x^i_α) in FU, $\widetilde{G}$ is given by

$$(2.17) \quad \begin{cases} \widetilde{G} = \sum\limits_{\alpha=1}^{n} \Big[\, x^a_\alpha u_{ij} \left(\psi^i_k \dfrac{\partial \psi^j_b}{\partial x^a} + \psi^j_b \dfrac{\partial \psi^i_k}{\partial x^a} \right) dx^k \otimes dx^b \\ \qquad\qquad + u_{ij} \psi^i_k \psi^j_b \, dx^k \otimes dx^b_\alpha + u_{ij} \psi^i_b \psi^j_k \, dx^b_\alpha \otimes dx^k \Big] . \end{cases}$$

Now, taking into account that $G_{kb} = u_{ij}\psi^i_k\psi^j_b$, it follows

$$\frac{\partial G_{kb}}{\partial x^a} = u_{ij} \left(\psi^i_k \frac{\partial \psi^j_b}{\partial x^a} + \psi^j_b \frac{\partial \psi^i_k}{\partial x^a} \right) ,$$

and then, combined with (2.17), we finally get

$$(2.18) \quad \begin{cases} \widetilde{G} = \sum\limits_{\alpha=1}^{n} \Big(x^k_\alpha \dfrac{\partial G_{ij}}{\partial x^k} \, dx^i \otimes dx^j \\ \qquad\qquad + G_{ij}\, dx^i \otimes dx^j_\alpha + G_{ij}\, dx^i_\alpha \otimes dx^j \Big) . \end{cases}$$

Remark 2.4.11 Because of the analogy with the terminology used by Mok in [65], $\widetilde{G}$ as given by (2.18) will be called *the complete lift of* G *to* FM and will be denoted by G^C (see [12,18]); in Section 3.2, in a different but equivalent context, we shall obtain the expression for the complete lift of tensor fields of type $(0, s)$, $s \geq 0$.

Linear groups

Let $V = \mathbf{R}^n$ and W a vector subspace of V, $dim\, W = k$. We denote by $Gl(V, W)$ the group of all $a \in Gl(V)$ such that $a(W) = W$.

$J^1_n W$ is canonically isomorphic to a vector subspace of $J^1_n V$ of dimension $(n+1)k$, so we can consider $Gl(J^1_n V, J^1_n W) \subset Gl(n+n^2, \mathbf{R})$. Without loss of generality, let us assume that, if $\{e_1, \ldots, e_n\}$ is the canonical basis of V, then W is generated by $\{e_1, \ldots, e_k\}$. Thus $G = Gl(V, W)$ is canonically isomorphic to the group of matrices $a \in Gl(n, \mathbf{R})$ of the form

$$a = \begin{pmatrix} a_0 & 0 \\ 0 & 0 \end{pmatrix} \, , \quad a_0 \in Gl(k, \mathbf{R}) \, ,$$

and its Lie algebra $\mathfrak{g} = gl(V, W)$ is isomorphic to the Lie algebra of matrices $B \in gl(n, \mathbf{R})$ of the form

$$B = \begin{pmatrix} B_0 & 0 \\ 0 & 0 \end{pmatrix} \, , \quad B_0 \in gl(k, \mathbf{R}) \, .$$

On the other hand, taking into account the isomorphism given in (1.10), $J^1_n W$ can be identified with the vector subspace of $\mathbf{R}^{n+n^2}$ of all points of the form

$$\begin{pmatrix} x^i \\ x^i_\alpha \end{pmatrix} \quad \text{with } x^i = x^i_\alpha = 0 \ \text{ for } k+1 \leq i \leq n \text{ and } 1 \leq \alpha \leq n \, .$$

Then, the following lemma is immediate.

Lemma 2.4.12 *Let $G = Gl(V, W)$ and $\tilde{G} = j_n(J^1_n G)$. Then*

$$\tilde{G} \subset Gl(J^1_n V, J^1_n W) \, .$$

Proposition 2.4.13 *If a manifold M possesses a k-dimensional differential system, then FM possesses a canonically induced $(n+1)k$-dimensional differential system. Moreover, if the differential system on M is completely integrable, then the differential system induced on FM is completely integrable too.*

Proof. M possesses a k-dimensional differential system if and only if M has a $Gl(V, W)$-structure with $dim\, V = dim\, M$. If M has a $Gl(V, W)$-structure, then FM has a $Gl(J^1_n V, J^1_n W)$-structure, by virtue of Lemma 2.4.12; since $dim\, J^1_n W = (n+1)k$, FM possesses a canonical $(n+1)k$-dimensional differential system.

On the other hand, it is well known that a differential system is completely integrable if and only if the associated $Gl(V, W)$-structure on M is integrable. Therefore, the last assertion follows from Theorem 2.3.7.□

Next, let $Sl(n, \mathbf{R})$ denote the special linear group, that is the group of all $a \in Gl(n, \mathbf{R})$ with $det\,(a) = 1$.

Lemma 2.4.14 *Let* $G = Sl(n, \mathbf{R})$ *and* $\tilde{G} = j_n(J_n^1 G)$. *Then*

$$\tilde{G} \subset Sl(n + n^2, \mathbf{R}) \ .$$

Proof. Direct from (1.12).□

The following proposition is now easy.

Proposition 2.4.15 *If* M *possesses an* $Sl(n, \mathbf{R})$*-structure, then* FM *possesses a canonical* $Sl(n + n^2, \mathbf{R})$*-structure.*

Chapter 3

Vector–Valued Differential Forms

Introduction

We propose to build on the development in Sections 1.3 and 1.4 concerning the object J^1_pV for a vector space V. As a vector space, it is isomorphic to $(p+1)$ copies of V and it induces a canonical embedding of Lie groups (cf. 1.4): $j_p: J^1_pGl(V) \longrightarrow Gl(J^1_pV)$. This embedding allows a lifting of linear representation of a Lie group G in V by the functor J^1_p. As candidates for V, we are particularly interested in tensor products of $\mathbf{R}^n$ and its dual; these being the fibre types of the geometrically interesting tangent tensor bundles on an n–dimensional manifold. A V–valued function on a manifold M is a section of $M \times V$; a V–valued r–form on M is a section of $\Lambda^r M \otimes TV$. We see how these are lifted by J^1_p. Similarly, we consider the lifting of V–valued functions on FM and their associated G–structures.

3.1 General Theory

We shall begin by presenting some general facts which will be used later on.

Let $\rho: G \longrightarrow Gl(V)$ be a linear representation of a Lie group G into V; then

$$\rho_1 = j_p \circ \rho^1 : J^1_pG \longrightarrow Gl(J^1_pV)$$

is again a linear representation which will be said to be induced by ρ. In fact

$$\rho_1(j^1\eta) = j_p(j^1(\rho \circ \eta)) \ , \ j^1\eta \in J^1_pG \ . \tag{3.1}$$

Denote the action of G on V by

$$\begin{aligned} \phi: G \times V &\longrightarrow V \\ (g, v) &\longmapsto \rho(g)(v) \end{aligned}$$

and let the map induced by J_p^1 be

$$\phi^1 : J_p^1 G \times J_p^1 V \longrightarrow J_p^1 V \ .$$

Lemma 3.1.1 *ϕ^1 is the action induced by ρ_1.*

Proof. Let $\lambda : Gl(V) \times V \longrightarrow V$, $\tilde{\lambda} : Gl(J_p^1 V) \times J_p^1 V \longrightarrow J_p^1 V$ be the natural actions. Then $\phi = \lambda \circ (\rho \times 1_V)$ and $\tilde{\phi} = \tilde{\lambda} \circ (\rho_1 \times 1_{J_p^1 V})$ are the actions induced by ρ and ρ_1, respectively.

Since $\lambda^1 = \lambda \circ (j_p \times 1_{J_p^1 V})$, it follows

$$\begin{aligned} \phi^1 &= (\lambda \circ (\rho \times 1_V))^1 = \lambda^1 \circ (\rho^1 \times 1_{J_p^1 V}) \\ &= \tilde{\lambda} \circ ((j_p \circ \rho^1) \times 1_{J_p^1 V}) = \tilde{\lambda} \circ (\rho_1 \times 1_{J_p^1 V}) \\ &= \tilde{\phi} \end{aligned}$$

and the lemma is proved.□

The following lemma will be useful later.

Lemma 3.1.2 *Let $\eta : \mathbf{R}^p \longrightarrow G$ be a differentiable map, $\tilde{g} = j^1 \eta \in J_p^1 G$, and $f : \mathbf{R}^p \longrightarrow TV$ such that $f(\mathbf{R}^p) \subset T_0 V$, the tangent space to V at 0. Define*

$$T\rho(\eta^{-1}) * f : \mathbf{R}^p \longrightarrow TV$$

by $(T\rho(\eta^{-1}) * f)(t) = T\rho(\eta^{-1}(t))(f(t))$. Then:

(i): $\alpha_V^{1,p}(j^1 f) \in T_0 J_p^1 V$;

(ii): $\alpha_V^{1,p}(j^1(T\rho(\eta^{-1}) * f)) = T\rho_1(\tilde{g})(\alpha_V^{1,p}(j^1 f))$.

Proof. Let $\sigma : \mathbf{R}^p \times \mathbf{R} \longrightarrow V$ be such that $f(t) = \dot{\sigma}_t$ for sufficiently small $t \in \mathbf{R}^p$, where $\sigma_t(u) = \sigma_u(t) = \sigma(t,u)$, $u \in \mathbf{R}$. Since $f(t) \in T_0 V$ for every t, then $\sigma_t(0) = 0 \in V$; consequently, $\sigma_0(t) = \sigma_t(0) = 0$, that is $\sigma_0 : \mathbf{R}^p \longrightarrow V$ is the constant map into $0 \in V$, and therefore $j^1 \sigma^0 = 0 \in J_p^1 V$.

But $\alpha_V^{1,p}(j^1 f) = \dot{\Sigma}$, $\Sigma : \mathbf{R} \longrightarrow J_p^1 V$ being given by $\Sigma(u) = j^1 \sigma^u$, and then $\Sigma(0) = j^1 \sigma^0$ and **(i)** is proved.

Now, $(T\rho(\eta^{-1}) * f)(t) = \dot{\sigma}'_t$, where $\sigma'_t(u) = \rho(\eta^{-1}(t))(\sigma_t(u))$; then, if we define

$$\Sigma' : \mathbf{R} \to J_p^1 V$$

by $\Sigma'(u) = j^1 \sigma'^u$, we get

$$\alpha_V^{1,p}(j^1(T\rho(\eta^{-1}) * f)) = \dot{\Sigma}' \ .$$

On the other hand, let $\rho(\eta^{-1}) * \sigma^u : \mathbf{R}^p \longrightarrow V$ be given by

$$(\rho(\eta^{-1}) * \sigma^u)(t) = \rho(\eta^{-1}(t))(\sigma^u(t)) ;$$

then, by virtue of (3.1),

$$\Sigma'(u) = j^1(\rho(\eta^{-1}) * \sigma^u) = j_p(j^1(\rho \circ \eta^{-1}))(j^1\sigma^u) = \rho_1(\tilde{g}^{-1})(j^1\sigma^u)$$

and (ii) is proved.□

Remark 3.1.3 Note that, as has been proved in Proposition 2.1.1 for the case of principal fibre bundles, if $E(M, \pi, V)$ is a vector bundle with standard fibre V, then $J^1_pE(J^1_pM, \pi^1, J^1_pV)$ is a vector bundle with standard fibre J^1_pV. This correspondence is functorial and preserves Whitney sums, that is, if $E'(M, \pi', V')$ is another vector bundle, then $J^1_p(E \oplus E')$ and $J^1_pE \oplus J^1_pE'$ are isomorphic vector bundles; in particular, for any integer $r \geq 2$, $J^1_p(\oplus_r TM)$ and $\oplus_r(J^1_pTM)$ are isomorphic vector bundles.

Particular cases

Now, let us consider the following vector spaces:

$$\begin{aligned} J^1_p\mathbf{R}^n &\simeq \mathbf{R}^n \times (\mathbf{R}^n)^p \equiv \mathbf{R}^{n+pn} , \\ \otimes^1_s\mathbf{R}^n &= \mathbf{R}^n \otimes (\mathbf{R}^n)^* \otimes \overset{s}{\cdots} \otimes (\mathbf{R}^n)^* \simeq L_s(\mathbf{R}^n, \mathbf{R}^n) , \\ \otimes_s\mathbf{R}^n &= (\mathbf{R}^n)^* \otimes \overset{s}{\cdots} \otimes (\mathbf{R}^n)^* \simeq L_s(\mathbf{R}^n, \mathbf{R}) , \end{aligned}$$

$L_s(V, W)$ denoting the vector space of s-linear maps of $V \times \overset{s}{\cdots} \times V$ into W; define $\pi: J^1_p\mathbf{R} \longrightarrow \mathbf{R}$ by

$$\pi(t, c_1, \ldots, c_p) = \sum_{\alpha=1}^{p} c_\alpha , \tag{3.2}$$

(compare with the map used in 2.4) and denote by $\{e^i\}$ the basis of $(\mathbf{R}^n)^*$ dual to the canonical basis $\{e_i\}$ of $\mathbf{R}^n$, $\{E^i, E^i_\alpha\}$ the basis of $(J^1_p\mathbf{R}^n)^*$ dual to the induced basis $\{E_i, E_{i_\alpha}\}$ of $J^1_p\mathbf{R}^n \equiv \mathbf{R}^{n+pn}$, and by

$$\{e_i \otimes e^{j_1} \otimes \ldots \otimes e^{j_s}, (e_i \otimes e^{j_1} \otimes \ldots \otimes e^{j_s})_\alpha\}$$

and

$$\{e^{j_1} \otimes \ldots \otimes e^{j_s}, (e^{j_1} \otimes \ldots \otimes e^{j_s})_\alpha\}$$

the induced basis of $J^1_p(\otimes^1_s\mathbf{R}^n)$ and $J^1_p(\otimes_s\mathbf{R}^n)$, respectively.

Then, the following homomorphisms of vector spaces can be defined:

(1): $i: J_p^1(\otimes_s^1 \mathbf{R}^n) \longrightarrow \otimes_s^1(J_p^1 \mathbf{R}^n) \equiv \otimes_s^1 \mathbf{R}^{n+pn}$

Let $j^1 f \in J_p^1(\otimes_s^1 \mathbf{R}^n)$ with $f: \mathbf{R}^p \to \otimes_s^1 \mathbf{R}^n$; then, $f(t) \in L_s(\mathbf{R}^n, \mathbf{R}^n)$ for each t.

We define

$$i(j^1 f) \in L_s(J_p^1 \mathbf{R}^n, J_p^1 \mathbf{R}^n)$$

as follows: for all $j^1 g_1, \ldots, j^1 g_s \in J_p^1 \mathbf{R}^n$,

$$i(j^1 f)(j^1 g_1, \ldots, j^1 g_s) = j^1(f * (g_1, \ldots, g_s))$$

where $(f * (g_1, \ldots, g_s))(t) = f(t)(g_1(t), \ldots, g_s(t))$. Then, a straightforward computation from the definitions leads to:

$$(3.3) \qquad \begin{cases} i(e_i \otimes e^{j_1} \otimes \cdots \otimes e^{j_s}) = E_i \otimes E^{j_1} \otimes \cdots \otimes E^{j_s} \\ \qquad + \sum\limits_{k=1}^{s} \sum\limits_{\alpha=1}^{p} E_{i_\alpha} \otimes E^{j_1} \otimes \cdots \otimes E_\alpha^{j_k} \otimes \cdots \otimes E^{j_s}; \\ i((e_i \otimes e^{j_1} \otimes \cdots \otimes e^{j_s})_\alpha) = E_{i_\alpha} \otimes E^{j_1} \otimes \cdots \otimes E^{j_s} \ . \end{cases}$$

(2): $i: J_p^1(\otimes_s \mathbf{R}^n) \to \otimes_s(J_p^1 \mathbf{R}^n) \equiv \otimes_s \mathbf{R}^{n+pn}$

Let $j^1 f \in J_p^1(\otimes_s \mathbf{R}^n)$ with $f: \mathbf{R}^p \to \otimes_s \mathbf{R}^n$; then, $f(t) \in L_s(\mathbf{R}^n, \mathbf{R})$, for each t. We define

$$i(j^1 f) \in L_s(\mathbf{R}^n, \mathbf{R})$$

as follows: for any $j^1 g_1, \ldots, j^1 g_s \in J_p^1 \mathbf{R}^n$,

$$i(j^1 f)(j^1 g_1, \ldots, j^1 g_s) = \pi(j^1 < f, (g_1, \ldots, g_s) >) \ ,$$

where $< f, (g_1, \ldots, g_s) > (t) = f(t)(g_1(t), \ldots, g_s(t))$. Then, a straightforward computation from the definitions leads to

$$(3.4) \qquad \begin{cases} i(e^{j_1} \otimes \cdots \otimes e^{j_s}) = \sum\limits_{k=1}^{s} \sum\limits_{\alpha=1}^{p} E^{j_1} \otimes \cdots \otimes E_\alpha^{j_k} \otimes \cdots \otimes E^{j_s} \ ; \\ i((e^{j_1} \otimes \cdots \otimes e^{j_s})_\alpha) = E^{j_1} \otimes \cdots \otimes E^{j_s} \ . \end{cases}$$

Both homomorphisms i above behave appropriately with respect to the canonical representations. To see this let

$$\rho: Gl(n, \mathbf{R}) \longrightarrow Gl(\otimes_s^1 \mathbf{R}^n) \quad \text{or} \quad \rho: Gl(n, \mathbf{R}) \longrightarrow Gl(\otimes_s \mathbf{R}^n)$$

be the canonical linear representations given by

$$(\rho(\eta) f)(\xi_1, \ldots, \xi_s) = \eta(f(\eta^{-1}\xi_1, \ldots, \eta^{-1}\xi_s)) \ , \ f \in \otimes_s^1 \mathbf{R}^n \ ,$$

or

$$(\rho(\eta)f)(\xi_1,\ldots,\xi_s) = f(\eta\xi_1,\ldots,\eta\xi_s) \ , \ f \in \otimes_s\mathbf{R}^n \ ,$$

for any $\eta \in Gl(n,\mathbf{R})$, $\xi_1,\ldots,\xi_s \in \mathbf{R}^n$ and let

$$\tilde{\rho}\colon J^1_pGl(n+pn,\mathbf{R}) \longrightarrow Gl(\otimes^1_s\mathbf{R}^{n+pn})$$

or

$$\tilde{\rho}\colon Gl(n+pn,\mathbf{R}) \longrightarrow Gl(\otimes_s\mathbf{R}^{n+pn})$$

be the analogous ones with

$$\rho_1\colon J^1_pGl(n,\mathbf{R}) \to Gl(J^1_p(\otimes^1_s\mathbf{R}^n)) \quad \text{or} \quad \rho_1\colon Gl(n,\mathbf{R}) \to Gl(J^1_p(\otimes_s\mathbf{R}^n))$$

the induced representations. Then, for any $\tilde{g} \in J^1_pGl(n,\mathbf{R})$,

$$i \circ \rho_1(\tilde{g}) = \tilde{\rho}(j_p(\tilde{g})) \circ i \ . \tag{3.5}$$

In fact, take $\eta\colon \mathbf{R}^p \longrightarrow Gl(n,\mathbf{R})$ with $\tilde{g} = j^1\eta$; then for arbitrary $j^1f \in J^1_p(\otimes^1_s\mathbf{R}^n)$,

$$\rho_1(\tilde{g})(j^1f) = j_p(j^1(\rho\circ\eta))(j^1f) = j^1((\rho\circ\eta) * f) \ ,$$

where $((\rho\circ\eta) * f)(t) = \rho(\eta(t))(f(t))$. Therefore,

$$i(\rho_1(\tilde{g})(j^1f))(j^1g_1,\ldots,j^1g_s) = j^1(((\rho\circ\eta) * f) * (g_1,\ldots,g_s)) \ ,$$

where

$$(((\rho\circ\eta) * f) * (g_1,\ldots,g_s))(t) = \eta(t)(f(t)(\eta(t)^{-1}g_1(t),\ldots,\eta(t)^{-1}g_s(t)))$$

for any $g_1,\ldots,g_s \in J^1_p\mathbf{R}^n$. On the other hand

$$\begin{aligned}
&\tilde{\rho}\,(j_p(\tilde{g})(i(j^1f))(j^1g_1,\ldots,j^1g_s) \\
&\quad = j_p(\tilde{g})(i(j^1f)(j_p(\tilde{g}^{-1})(j^1g_1),\ldots,j_p(\tilde{g}^{-1})(j^1g_s))) \quad \text{(by Def. 1.4.1)} \\
&\quad = j_p(\tilde{g})(i(j^1f)(j^1(\eta^{-1} * g_1),\ldots,j^1(\eta^{-1} * g_s))) \\
&\quad = j_p(\tilde{g})(j^1(f * (\eta^{-1} * g_1,\ldots,\eta^{-1} * g_s))) \quad \text{(by Def. 1.4.1)} \\
&\quad = j^1(\eta * (f * (\eta^{-1}g_1,\ldots,\eta^{-1}g_s))) \ ,
\end{aligned}$$

and (3.5) holds for the type $(1,s)$. We omit the proof for the type $(0,s)$, which is similar.

3.2 Applications

Let V be a vector space as before. Define injections

$$i_0, i_\alpha : V \longrightarrow J^1_p V \ , \ 1 \leq \alpha \leq p \ ,$$

by

$$i_0(\xi) = j^1 f_\xi \ , \ i_\alpha(\xi) = j^1 f_{\alpha,\xi} \ , \ \xi \in V \ ,$$

where $f_\xi, f_{\alpha,\xi} : \mathbf{R}^p \longrightarrow V$ are given by

$$f_\xi(t) = \xi \ , \ f_{\alpha,\xi}(t) = t^\alpha \xi \ , \ t = (t^1, \ldots, t^p) \in \mathbf{R}^p \ .$$

Prolongation of functions and forms

Let $h: M \longrightarrow V$ be a V-valued differentiable function, and let

$$h^1 : J^1_p M \longrightarrow J^1_p V$$

be the induced one; define

$$h^{(\alpha)} = i_\alpha \circ h \circ \pi_M : J^1_p M \longrightarrow J^1_p V \ , \ \alpha \in \{0, 1, 2, ..., p\} \ .$$

If h is locally expressed by $h(x^i) = h^a(x^i) e_a$ with respect to the basis $\{e_\alpha\}$, then the local expressions of h^1 and $h^{(\alpha)}$ with respect to the induced basis $\{E_a, E_{a_\alpha}\}$ are:

$$\begin{aligned} h^1(x^i, x^i_\alpha) &= h^a(x^i) E_a + x^j_\alpha \frac{\partial h^a}{\partial x^j} E_{a_\alpha} \ , \\ h^{(0)}(x^i, x^i_\alpha) &= h^a(x^i) E_a \ ; \\ h^{(\alpha)}(x^i, x^i_\alpha) &= h^a(x^i) E_{a_\alpha} \ , \ 1 \leq \alpha \leq p \ . \end{aligned}$$

Hence, recalling (1.3) and (1.4), for any vector field X on M,

$$\begin{aligned} &X^C h^1 = (Xh)^1 \ , \qquad X^C h^{(\alpha)} = (Xh)^{(\alpha)} \ , \ 0 \leq \alpha \leq p \ , \\ &X^{(\alpha)} h^1 = (Xh)^{(\alpha)} \ , \quad X^{(\alpha)} h^{(\beta)} = 0 \ , \ 1 \leq \alpha \leq p, \ 0 \leq \beta \leq p \ . \end{aligned}$$

In particular, if $V = \mathfrak{g}$ is a Lie algebra, $f, g : M \to \mathfrak{g}$ arbitrary maps, the bracket product map $[f, g] : M \to \mathfrak{g}$ is defined by $[f, g](x) = [f(x), g(x)]$, $x \in M$. Then, the following identities hold

$$(3.6) \qquad \begin{cases} [f, g]^1 = [f^1, g^1] \ , \ [f, g]^{(0)} = [f^{(0)}, g^{(0)}] \ , \\ [f, g]^{(\alpha)} = [f^{(\alpha)}, g^1] = [f^1, g^{(\alpha)}] = [f^{(0)}, g^{(\alpha)}] = [f^{(\alpha)}, g^{(0)}] \ , \\ [g^{(\alpha)}, g^{(\beta)}] = 0 \ , \ 1 \leq \alpha, \beta \leq p \ . \end{cases}$$

Let G be a Lie group acting on M on the right through the differentiable map $\psi: M \times G \longrightarrow M$, $\rho: G \longrightarrow Gl(V)$ a linear representation and $\phi: G \times V \longrightarrow V$ the action induced by ρ. For each $h: M \longrightarrow V$, let

$$\begin{aligned} i \times h: M \times G &\longrightarrow G \times V \\ (x, g) &\longmapsto (g^{-1}, h(x)) \ . \end{aligned}$$

Definition 3.2.1 *A differentiable map $h: M \longrightarrow V$ is called of type (ρ, V) if $h \circ \phi = \phi \circ (i \times h)$, or equivalently if $h(xg) = \rho(g^{-1})(h(x))$ for $x \in M$, $g \in G$.*

Theorem 3.2.2 *If the differentiable map $h: M \longrightarrow V$ is of type (ρ, V), then the induced map $h^1: J^1_p M \to J^1_p V$ is of type $(\rho_1, J^1_p V)$.*

Proof. $J^1_p G$ acts on $J^1_p M$ on the right through ψ^1 (recall 1.2), and ϕ^1 is the action of $J^1_p G$ on $J^1_p V$ induced by $\rho_1: J^1_p G \longrightarrow Gl(J^1_p V)$ (recall Lemma 3.1.1). On the other hand, $(i \times h)^1 = i \times h^1$, hence $h^1 \circ \psi^1 = (h \circ \psi)^1 = (\phi \circ (i \times h))^1 = \phi^1 \circ (i \times h)^1$, which proves the theorem.□

Next, let ω be a V-valued 1-form on M; ω can be considered canonically as a differentiable map $\omega: TM \longrightarrow TV$ which is a linear map of the tangent space $T_x M$ with values in the tangent space $T_0 V$ for each $x \in M$.

So, let $\omega: TM \longrightarrow TV$ be a V-valued 1-form on M, and define a differentiable map

$$\omega_1: TJ^1_p M \longrightarrow TJ^1_p V$$

by setting

$$\omega_1 = \alpha_V^{1,p} \circ \omega^1 \circ \alpha_M^{p,1} \ , \tag{3.7}$$

where $\omega^1: J^1_p TM \longrightarrow J^1_p TV$ is the map induced by ω.

Let (U, x^i) be a chart on M, $(TJ^1_p U, (x^i, x^i_\alpha; \dot{x}^i, \dot{x}^i_\alpha))$ the induced chart on $TJ^1_p M$; if ω is given in TU by

$$\omega(x^i; \dot{x}^i) = (0; \omega^a_k(x^i)\dot{x}^k)$$

or, equivalently, if ω appears in U as

$$\omega = (\omega^a_k \, dx^k) \, e_a \ ,$$

then a routine computation, using 1.1 and (1.5), leads to the following expression for ω_1 in $J^1_p U$:

$$\omega_1 = (\omega^a_k \, dx^k) \, E_a + \left(x^j_\alpha \frac{\partial \omega^a_k}{\partial x^j} \, dx^j + \omega^a_k \, dx^k_\alpha \right) E_{a_\alpha} \ ,$$

which implies that ω_1 is a (well defined) J^1_pV-valued 1-form on J^1_pM. Moreover, one easily obtains the following identities, which can be considered as an alternative definition of ω_1:

$$\tag{3.8} \begin{cases} \omega_1(X^C) &= (\omega(X))^1 , \\ \omega_1(X^{(\alpha)}) &= (\omega(X))^{(\alpha)} , \; 1 \leq \alpha \leq p , \end{cases}$$

for every vector field X on M.

Definition 3.2.3 *ω_1 given by (3.7), or by (3.8), will be called the prolongation of ω to J^1_pM.*

This definition can be extended to higher degrees as follows. To see this let ω be a V-valued r-form on M, $r \geq 2$, considered as a differentiable map

$$\omega: \oplus_r TM \longrightarrow TV$$

which is a linear skew-symmetric map of $\oplus_r T_xM$ with values in T_0V for each $x \in M$. We define the prolongation ω_1 of ω to J^1_pM as the map

$$\omega_1: \oplus_r TJ^1_pM \longrightarrow TJ^1_pV$$

given by

$$\omega_1 = \alpha_V^{1,p} \circ \omega^1 \circ \simeq \circ (\oplus_r \alpha_M^{p,1}) ,$$

where $\simeq: \oplus_r(J^1_pTM) \longrightarrow J^1_p(\oplus_r TM)$ is the canonical isomorphism of vector bundles (see Remark 3.1.3). Thus, if ω is locally expressed on M by

$$\omega = (\omega^a_{i_1\cdots i_r}\, dx^{i_1} \wedge \ldots \wedge dx^{i_r})\, e_a ,$$

then ω_1 is locally expressed on J^1_pM by

$$\begin{aligned} \omega_1 = \;& (\omega^a_{i_1\cdots i_r}\, dx^{i_1} \wedge \cdots \wedge dx^{i_r})\, E_a \\ &+ \Big(x^j_\alpha \frac{\partial \omega_{i_1\cdots i_r}}{\partial x^j}\, dx^{i_1} \wedge \cdots \wedge dx^{i_r} \\ &\quad + \sum_{k=1}^{r} \omega^a_{i_1\cdots i_r}\, dx^{i_1} \wedge \cdots \wedge dx^{i_k}_\alpha \wedge \cdots \wedge dx^{i_r} \Big) E_{a_\alpha} . \end{aligned}$$

Therefore, for any vector fields $X_1, \ldots, X_r$ on M

$$\tag{3.9} \begin{cases} \omega_1(X_1^C, \ldots, X_r^C) = (\omega(X_1, \ldots, X_r))^1 ; \\ \omega_1(X_1^C, \ldots, X_j^{(\alpha)}, \ldots, X_r^C) = (\omega(X_1, \ldots, X_r))^{(\alpha)} ; \\ \omega_1(X_1^C, \ldots, X_i^{(\alpha)}, \ldots, X_j^{(\beta)}, \ldots, X_r^C) = 0 , \; 1 \leq \alpha, \beta \leq p . \end{cases}$$

identities which, in fact, can be considered as an alternative definition of ω_1.

The following identities are easy to prove:

$$(3.10) \qquad \begin{cases} (\omega+\tau)_1 &= \omega_1+\tau_1 \ , \\ (\lambda\omega)_1 &= \lambda\omega_1 \ (\lambda \in \mathbf{R}) \ , \\ \iota_{X^C}\omega_1 &= (\iota_X\omega)_1 \ , \\ \mathcal{L}_{X^C}\omega_1 &= (\mathcal{L}_X\omega)_1 \ , \\ d\omega_1 &= (d\omega)_1 \ . \end{cases}$$

Moreover, if $\phi: M \longrightarrow N$ is a differentiable map, then for any V-valued r-form ω on N,

$$(3.11) \qquad (\phi^*\omega)_1 = (\phi^1)^*\omega_1 \ .$$

For, by definition,

$$(\phi^*\omega)_1 = \alpha_V^{1,p} \circ \omega^1 \circ (\oplus_r T\phi)^1 \circ \simeq \circ (\oplus_r \alpha_M^{p,1}) \ ;$$

but $(\oplus_r T\phi)^1 \circ \simeq \ = \ \simeq \circ(\oplus_r (T\phi)^1)$ and then, using the results in 1.1 it follows

$$\begin{aligned} (\phi^*\omega)_1 &= \alpha_V^{1,p} \circ \omega^1 \circ \simeq \circ (\oplus_r (T\phi)^1) \circ (\oplus_r \alpha_M^{p,1}) \\ &= \alpha_V^{1,p} \circ \omega^1 \circ \simeq \circ (\oplus_r \alpha_M^{p,1}) \circ (\oplus_r T\phi^1) \\ &= \omega_1 \circ (\oplus_r T\phi)^1 \\ &= (\phi^1)^*\omega_1 \ . \end{aligned}$$

Let be $\psi: M \times G \longrightarrow M$, $\rho: G \longrightarrow Gl(V)$ and $\phi: G \times V \longrightarrow V$ as before, and for each $g \in G$, let $R_g: M \longrightarrow M$ be given by $R_g(x) = \psi(x,g) = xg$.

Definition 3.2.4 *A V-valued r-form $\omega: \oplus_r TM \longrightarrow TV$ is said to be of type (ρ, V) if, for any $g \in G$, $\omega \circ (\oplus_r TR_g) = T\rho(g^{-1}) \circ \omega$.*

Theorem 3.2.5 *Let ω be a V-valued r-form on M of type (ρ, V). Then, the prolongation ω_1 of ω to $J_p^1 M$ is an r-form of type $(\rho_1, J_p^1 V)$, that is, for any $\tilde{g} \in J_p^1 G$*

$$\omega_1 \circ (\oplus_r TR_{\tilde{g}}) = T\rho_1(\tilde{g}^{-1}) \circ \omega_1 \ .$$

Proof. The extension for $r \geq 2$ being clear, we shall prove the theorem only for $r = 1$.

Let $\tilde{g} \in J_p^1 G$, $v \in T_{\tilde{x}} J_p^1 M$, $\tilde{x} \in J_p^1 M$, and

$$\eta: \mathbf{R}^p \longrightarrow G \quad , \quad \phi: \mathbf{R} \longrightarrow J_p^1 M$$

such that $\tilde{g} = j^1\eta$, $\phi(0) = \tilde{x}$ and $\dot{\phi} = v$. Then, there exists a differentiable map $\psi: \mathbf{R} \times \mathbf{R}^p \longrightarrow M$ and $\delta > 0$ such that $\phi(t) = j^1\psi_t$ for $|t| < \delta$, where $\psi_t(u) = \psi^u(t) = \psi(t,u)$, $t \in \mathbf{R}$, $u \in \mathbf{R}^p$.

Therefore, for small t,

$$(R_{\tilde{g}} \circ \phi)(t) = j^1\psi'_t \ ,$$

where $\psi'_t(u) = \psi'(t,u) = \psi(t,u)\eta(u)$. Define $\Psi': \mathbf{R}^p \to TM$ by setting $\Psi'(u) = \dot{\psi}'^u$; then

$$TR_{\tilde{g}}(v) = \overset{\bullet}{\overline{(R_{\tilde{g}} \circ \phi)}} \ ,$$

and, by definition of $\alpha_M^{p,1}$,

$$\alpha_M^{p,1}(TR_{\tilde{g}}(v)) = j^1\Psi' \ .$$

In particular, $\alpha_M^{p,1}(v) = j^1\Psi$, $\Psi: \mathbf{R}^p \longrightarrow TM$ being given by $\Psi(u) = \dot{\psi}^u$. Now

$$\begin{aligned}
(\omega \circ \Psi') &= \omega(\dot{\psi}'^u) \\
&= \omega(TR_{\eta(u)}(\dot{\psi}^u)) \quad \text{(because } \omega \text{ is of type } (\rho, V)) \\
&= T\rho(\eta^{-1}(u))(\omega(\dot{\psi}^u)) \\
&= T\rho(\eta^{-1}(u))((\omega \circ \Psi)(u)) \ .
\end{aligned}$$

Hence

$$(\omega^1 \circ \alpha_M^{p,1})(TR_{\tilde{g}}(v)) = \omega^1(j^1\Psi') = j^1(\omega \circ \Psi') = j^1(T\rho(\eta^{-1}) * (\omega \circ \Psi)) \ ,$$

and applying Lemma 3.1.2 for $f = \omega \circ \Psi$, we deduce

$$\begin{aligned}
(\alpha_V^{1,p} \circ \omega^1 \circ \alpha_M^{p,1})(TR_{\tilde{g}}(v)) &= \alpha_V^{1,p}(j^1(T\rho(\eta^{-1}) * (\omega \circ \Psi)) \\
&= T\rho_1(\tilde{g}^{-1})(\alpha_V^{1,p}(j^1(\omega \circ \Psi))) \ .
\end{aligned}$$

But $j^1(\omega \circ \Psi) = \omega^1(j^1\Psi) = \omega^1(\alpha_M^{p,1}(v)) = (\omega^1 \circ \alpha_M^{p,1})(v)$, and therefore

$$\omega_1(TR_{\tilde{g}}(v)) = T\rho_1(\tilde{g}^{-1})(\omega_1(v)) \ ,$$

which ends the proof.□

Remark 3.2.6 All the results in this section apply when $V = \mathfrak{g}$ is a Lie algebra. Nevertheless, the special significance of differential forms taking values in a Lie algebra, mainly in connection theory, makes suitable a slightly different approach which will be developed in Section 4.1.

Complete lift of functions and tensor fields

Let V, $\tilde{V}$ be vector spaces,

$$\rho: Gl(n, \mathbf{R}) \longrightarrow Gl(V)$$

and

$$\tilde{\rho}: Gl(n+n^2, \mathbf{R}) \longrightarrow Gl(\tilde{V})$$

linear representations, and $i: J_n^1 V \longrightarrow \tilde{V}$ a homomorphism such that

$$i \circ \rho_1(\tilde{g}) = \rho(j_n(\tilde{g})) \circ i \,, \quad \tilde{g} \in J_n^1 Gl(n, \mathbf{R}) \,. \tag{3.12}$$

Let $t: FM \longrightarrow V$ be a differentiable function of type (ρ, V); from Theorem 3.2.2 we know that the prolongation t^1 of t to $J_n^1 FM$ is a differentiable function of type $(\rho_1, J_n^1 V)$. Then, given $\tilde{y} \in FJ_n^1 M$, choose $\tilde{x} \in J_n^1 FM$ and $\tilde{a} \in Gl(n+n^2, \mathbf{R})$ such that $\tilde{y} = j_M(\tilde{x})\tilde{a}$, and define

$$\tilde{t}: FJ_n^1 M \longrightarrow \tilde{V}$$

by setting

$$\tilde{t}(\tilde{y}) = \tilde{\rho}(\tilde{a}^{-1})((i \circ t^1)(\tilde{x})) \,. \tag{3.13}$$

Identity (3.12) implies that $\tilde{t}(\tilde{y})$ does not depend on the choices of $\tilde{x}$ and $\tilde{a}$; moreover, $\tilde{t}$ is obviously of type $(\tilde{\rho}, \tilde{V})$ and, hence, we adopt the following definition.

Definition 3.2.7 *The differentiable function $\tilde{t}: FJ_n^1 M \longrightarrow \tilde{V}$ given by (3.13) will be called the complete lift of t to $FJ_n^1 M$.*

This procedure of "lifting" functions on FM to $FJ_n^1 M$ has some interesting applications when particular choices of V, $\tilde{V}$, ρ, $\tilde{\rho}$ and i are fixed. In order to give a clear description of such applications, let us firstly recall the one–to–one correspondence between tensor fields F on M of type (r, s) and differentiable functions $t: FM \longrightarrow \otimes_s^r \mathbf{R}^n$ of type $(\rho, \otimes_s^r \mathbf{R})$, ρ being the canonical representation of $Gl(n, \mathbf{R})$ in $\otimes_s^r \mathbf{R}^n$.

This correspondence is locally given as follows. Let (U, x^i) be a chart on M, (FU, x^i, X_j^i) the induced chart on FM (considered as a principal fibre bundle with structure group $Gl(n, \mathbf{R})$). Now take a tensor field F (resp. the function t) given in U (resp. in FU) by its local components $F_{i_1 \cdots i_s}^{j_1 \cdots j_r}$ (resp. $t_{i_1 \cdots i_s}^{j_1 \cdots j_r}$); then

$$t_{h_1 \cdots h_s}^{k_1 \cdots k_r}(x^i, X_j^i) = F_{i_1 \cdots i_s}^{j_1 \cdots j_r}(x^i) X_{h_1}^{i_1} \cdots X_{h_s}^{i_s} \overline{X}_{j_1}^{k_1} \cdots \overline{X}_{j_r}^{k_r} \,, \tag{3.14}$$

where $(\overline{X}_j^i) = (X_j^i)^{-1}$.

Next, observe that if (x^i, X^i_j) are the fibre coordinates on FU, then we obtain induced coordinates $(x^i, X^i_j, x^i_\alpha, X^i_{j\alpha})$ on $J^1_n FU$. Now, if (y^i, y^i_α, Y^A_B) are the natural coordinates on $FJ^1_n U$, then the canonical imbedding

$$j_M: J^1_n FM \to FJ^1_n M$$

is locally expressed by the equations:

$$(3.15) \qquad j_M : \begin{cases} y^i = x^i \,, & y^i_\alpha = x^i_\alpha \,, \\ Y^i_j = X^i_j \,, & Y^{i_\alpha}_j = X^i_{j\alpha} \,, \quad Y^i_{j_\alpha} = 0 \,, \\ Y^{i_\alpha}_{j_\beta} = \delta^\alpha_\beta X^i_j \,; \end{cases}$$

where $i_\alpha = \alpha n + i$.

Now, note that given a differentiable function $f: M \longrightarrow \mathbf{R}$, then

$$t = f \circ \pi_M: FM \longrightarrow \mathbf{R}$$

is of type $(\rho, \mathbf{R})$, $\rho: Gl(n, \mathbf{R}) \longrightarrow Gl(\mathbf{R})$ being the trivial representation. Conversely, if $t: FM \longrightarrow \mathbf{R}$ is of type $(\rho, \mathbf{R})$, then t projects down to a differentiable function $f: M \longrightarrow \mathbf{R}$.

So, take $V = \tilde{V} = \mathbf{R}$, ρ and $\tilde{\rho}$ being the trivial representations, and $t = \pi$ given by (3.2); then (3.12) holds trivially, and for any $f: M \longrightarrow \mathbf{R}$ we can consider the complete lift $\tilde{t}: FJ^1_n M \longrightarrow \mathbf{R}$ of $t = f \circ \pi$.

Thus, given $\tilde{y} \in FJ^1_n M$, we choose $\tilde{x} = (x^i, X^i_j, x^i_\alpha, X^i_{j\alpha}) \in J^1_n FM$ and $\tilde{a} \in Gl(n + n^2, \mathbf{R})$ such that $\tilde{y} = j_M(\tilde{x})\tilde{a}$; a straightforward computation leads to

$$\tilde{t}(\tilde{y}) = \sum_{\alpha=1}^{n} x^j_\alpha \frac{\partial f}{\partial x^j} \,,$$

and, then, the projection of $\tilde{t}$ to $J^1_n M$ is precisely the differentiable function $\tilde{f}: J^1_n M \to \mathbf{R}$ given by

$$(3.16) \qquad \tilde{f}(x^i, x^i_\alpha) = \sum_{\alpha=1}^{j} \frac{\partial f}{\partial x^j} \,.$$

The restriction $f^C = f_{|FM}$ is called *the complete lift of f to FM.*

The correspondence $f \longrightarrow f^C$ behaves, with respect to the ring structure of the space of differentiable functions on M, as follows:

$$(3.17) \qquad \begin{cases} (f+g)^C = f^C + g^C \,, \\ (fg)^C = f^C g^V + f^V g^C \,, \end{cases}$$

where $f^V = f \circ \pi_M$. On the other hand, for any vector field X on M,

$$X^C f^C = (Xf)^C \quad , \quad X^{(\alpha)} f^C = (Xf)^V = X^C f^V \ . \tag{3.18}$$

Next, take $V = \otimes^1_s \mathbf{R}^n$, $\tilde{V} = \otimes^1_s (J^1_n \mathbf{R}^n)$, ρ and $\tilde{\rho}$ be the canonical representations of $Gl(n, \mathbf{R})$ and $Gl(n+n^2, \mathbf{R})$ into V and $\tilde{V}$, respectively, and $i: J^1_n V \longrightarrow \tilde{V}$ the homomorphism given by (3.3) for $p = n$.

From (3.5) we know that (3.12) holds and hence, if F is a tensor field on M of type $(1, s)$ with associated function $f: FM \longrightarrow \otimes^1_s \mathbf{R}^n$ of type $(\rho, \otimes^1_s \mathbf{R}^n)$, then $\tilde{t}: FJ^1_n M \longrightarrow \otimes^1_s \mathbf{R}^{n+n^2}$ is well defined by (3.13) and determines a tensor field $\widetilde{F}$ on $J^1_n M$ of type $(1, s)$.

To compute $\widetilde{F}$, we assume $\tilde{y} = j_M(\tilde{x})$ with $\tilde{x} = (x^i, X^i_j, x^i_\alpha, X^i_{j_\alpha}) \in J^1_n FM$; then, using (3.15), (3.3), (3.13) and (3.14), a direct computation leads to

$$\begin{aligned}
\tilde{t}(\tilde{y}) &= F^i_{j_1 \cdots j_s} X^{j_1}_{k_1} \cdots X^{j_s}_{k_s} \overline{X}^h_i E_h \otimes E^{k_1} \otimes \cdots \otimes E^{k_s} \\
&+ \Big(x^k_\alpha X^{j_1}_{k_1} \cdots X^{j_s}_{k_s} \overline{X}^h_i (\partial_k F^i_{j_1 \cdots j_s}) - X^k_{i_\alpha} X^{j_1}_{k_1} \cdots X^{j_s}_{k_s} \overline{X}^h_k \overline{X}^l_i F^i_{j_1 \cdots j_s} \\
&\quad + \sum_{l=1}^{s} X^{j_l}_{k_{l\alpha}} X^{j_1}_{k_1} \cdots X^{j_{l-1}}_{k_{l-1}} X^{j_{l+1}}_{k_{l+1}} \cdots \\
&\qquad \cdots X^{j_s}_{k_s} \overline{X}^h_i F^i_{j_1 \cdots j_s} \Big) E_{h_\alpha} \otimes E^{k_1} \otimes \cdots \otimes E^{k_s} \\
&+ \sum_{l=1}^{s} \delta^\beta_\alpha X^{j_1}_{k_1} \cdots X^{j_s}_{k_s} \overline{X}^h_i F^i_{j_1 \cdots j_s} E_{h_\alpha} \otimes E^{k_1} \cdots \otimes E^{k_l}_\beta \otimes \cdots \otimes E^{k_s} \ ,
\end{aligned}$$

where $F^i_{j_1 \cdots j_s}$ are the local components of F and $\partial_k = (\partial / \partial x^k)$.

Then, using again (3.14) with the appropriate changes to $J^1_n M$ and $FJ^1_n M$, it follows

$$\left\{ \begin{aligned}
\widetilde{F}(x^i, x^i_\alpha) &= F^i_{j_1 \cdots j_s} \frac{\partial}{\partial x^i} \otimes dx^{j_1} \otimes \cdots \otimes dx^{j_s} \\
&+ (x^h_\alpha \partial_h F^i_{j_1 \cdots j_s}) \frac{\partial}{\partial x^i_\alpha} \otimes dx^{j_1} \otimes \cdots \otimes dx^{j_s} \\
&+ \sum_{k=1}^{s} \delta^\beta_\alpha F^i_{j_1 \cdots j_s} \frac{\partial}{\partial x^i_\alpha} \otimes dx^{j_1} \otimes \cdots \otimes dx^{j_k}_\beta \otimes \cdots \otimes dx^{j_s}
\end{aligned} \right. \tag{3.19}$$

The restriction $F^C = \widetilde{F}_{|FM}$ is called *the complete lift of F to FM*, and it has been defined in an equivalent way by K.P. Mok [65].

Note that if X is a vector field on M, that is for $s = 0$, then $\widetilde{X}$ is just the complete lift X^C of X to FM given by (1.3).

In [65] it is proved that F^C can be alternatively defined as the unique tensor field on FM of type $(1, s)$ such that

$$F^C(X^C_1, \ldots, X^C_s) = \{F(X_1, \ldots, X_s)\}^C \ , \tag{3.20}$$

for any vector fields $X_1, \ldots, X_s$ on M.

Obviously the correspondence $F \longrightarrow F^C$ is linear. More properties and applications are developed in the following Chapters.

Now, let be $V = \otimes_s \mathbf{R}^n$, $\widetilde{V} = \otimes_s(J^1_n\mathbf{R}^n) = \otimes_s\mathbf{R}^{n+n^2}$, ρ and $\tilde{\rho}$ the canonical representations of $Gl(n,\mathbf{R})$ and $Gl(n+n^2,\mathbf{R})$ in V and $\widetilde{V}$, respectively, and let $i: J^1_nV \longrightarrow \widetilde{V}$ be given by (3.4) for $p = n$.

Once more (3.12) holds in this case, and if F is a tensor field on M of type $(0,s)$ with associated function

$$t: FM \longrightarrow \otimes_s\mathbf{R}^n ,$$

then its complete lift $\tilde{t}$, given by (3.13), appears locally as follows. With $\tilde{y}$,$\tilde{x}$ as before,

$$\begin{aligned}\tilde{t}(\tilde{y}) = \sum_{\alpha=1}^{n}\Big((&x^h_\alpha X^{i_1}_{k_1}\cdots X^{i_s}_{k_s}(\partial_h F_{i_1\cdots i_s}) \\ &+ \sum_{l=1}^{s} X^{i_l}_{k_l\alpha} X^{i_1}_{k_1}\cdots X^{i_{l-1}}_{k_{l-1}} X^{i_{l+1}}_{k_{l+1}}\cdots X^{i_s}_{k_s} F_{i_1\ldots i_s})E^{k_1}\otimes\cdots\otimes E^{k_s} \\ &+ \sum_{l=1}^{s} X^{i_1}_{k_1}\cdots X^{i_s}_{k_s} F_{i_1\cdots i_s} E^{k_1}\otimes\cdots\otimes E^{k_l}_\alpha\otimes\cdots\otimes E^{k_s}\Big)\end{aligned}$$

where $F_{i_1\cdots i_s}$ are the local components of F. Hence, the local expression for $\widetilde{F}$ on J^1_nM is:

$$(3.21)\qquad \left\{\begin{aligned}\tilde{F}(x^i, x^i_\alpha) = \sum_{\alpha=1}^{n}\Big\{&x^h_\alpha \partial_h F_{i_1\cdots i_s}\, dx^{i_1}\otimes\cdots\otimes dx^{i_s} \\ &+ \sum_{k=1}^{s} F_{i_1\cdots i_s}\, dx^{i_1}\otimes\cdots\otimes dx^{i_k}_\alpha\otimes\cdots\otimes dx^{i_s}\Big\}\end{aligned}\right.$$

The restriction $F^C = \widetilde{F}_{|FM}$ is called *the complete lift of* F *to* FM. F^C can be defined in an equivalent way as the unique tensor field on FM of type $(0,s)$ such that (3.20) is also valid. We refer the reader to Chapter 8, where a detailed study of other properties and applications of this definition will be developed. Note that the following identities are satisfied:

$$(F+G)^C = F^C + G^C \ , \ (F\otimes G)^C = F^C\otimes G^V + F^V\otimes G^C \ ,$$

for any covariant tensor fields F, G on M.

Prolongation of G–structures

Let $u \in V$ be a fixed element of V, G_u the isotropy group of u with respect to the linear representation $\rho: Gl(n,\mathbf{R}) \longrightarrow Gl(V)$, and

$$V_u = \{\rho(g)u \mid g \in Gl(n,\mathbf{R})\} \ .$$

It is well known that there exists a one-to-one correspondence between G_u-structures P_{G_u} on M and differentiable functions $t: FM \longrightarrow V$ of type (ρ, V) such that: (i) $t(FM) \subset V_u$; (ii) t is a differentiable map of FM into V_u; this correspondence is given by setting $P_{G_u} = t^{-1}(u)$.

Let $\tilde{\rho}: Gl(n+n^2, \mathbf{R}) \longrightarrow Gl(\tilde{V})$ be another linear representation, and take $i: J^1_n V \longrightarrow \tilde{V}$ a homomorphism of vector spaces such that (3.12) holds. Let $u^1 \in J^1_n V$ be the 1-jet at 0 of the constant map of $\mathbf{R}^n$ into $u \in V$, and put $\tilde{u} = i(u^1)$. Then, from Definition 1.4.1, (3.1) and (3.12) it follows easily:

$$\overline{G_u} = j_n(J^1_n G_u) \subset G_{\tilde{u}} , \tag{3.22}$$

$G_{\tilde{u}}$ being the isotropy group of $\tilde{u}$.

Let $t: FM \longrightarrow V$ define the G_u-structure P_{G_u} on M, and let

$$\tilde{t}: FJ^1_n M \longrightarrow \tilde{V}$$

be the complete lift of t.

Lemma 3.2.8 $\tilde{t}(FJ^1_n M) \subset \tilde{V}_{\tilde{u}}$.

Proof. Firstly, note that $t^1(\tilde{x}) \in J^1_n V$ for any $\tilde{x} \in J^1_n FM$. Next, given an arbitrary $j^1 h \in J^1_n V_u$, define $\tilde{g} \in J^1_n Gl(n, \mathbf{R})$ by $\tilde{g} = j^1 \eta$, where $\eta: \mathbf{R}^n \to Gl(n, \mathbf{R})$ is determined by the identity $h(r) = \rho(\eta(r))u$, $r \in \mathbf{R}^n$. Then, from Definition 1.4.1 and (3.1), it follows $j^1 h = \rho_1(\tilde{g})(u^1)$; hence $i(j^1 h) = \tilde{\rho}(j_n(\tilde{g})(\tilde{u}))$, and the result follows from (3.13).□

Thus, $P_{G_{\tilde{u}}} = \tilde{t}^{-1}(\tilde{u})$ defines a $G_{\tilde{u}}$-structure on $J^1_n M$ and hence, by restriction, on FM; so, we can state:

Theorem 3.2.9 *If M admits a G_u-structure defined by a tensor t of type (ρ, V) and V_u-valued, then the complete lift $\tilde{t}$ of t defines a $G_{\tilde{u}}$-structure on FM.*

Remark 3.2.10 Theorem 3.2.9 improves some particular results stated in Section 2.4. In fact, let $\overline{P_{G_u}} = j_M(J^1_n P_{G_u})_{|FM}$ be the prolongation of P_{G_u} to FM; then $\overline{P_{G_u}} \subset P_{G_{\tilde{u}}|FM}$ and, hence, $P_{G_{\tilde{u}}|FM}$ is isomorphic to the canonical prolongation of $\tilde{P}_{G_u}$ induced by the injection $i: \overline{G_u} \longrightarrow G_{\tilde{u}}$.

Now, assume $V = \otimes^1_s \mathbf{R}^n$, $\tilde{V} = \otimes^1_s (J^1_n \mathbf{R}^n)$ for the type $(1, s)$, or $V = \otimes_s \mathbf{R}^n$, $\tilde{V} = \otimes_s (J^1_n \mathbf{R}^n) = \otimes_s \mathbf{R}^{n+n^2}$ for the type $(0, s)$, and let ρ and $\tilde{\rho}$ be the canonical representations of $Gl(n, \mathbf{R})$ and $Gl(n+n^2, \mathbf{R})$ in V and $\tilde{V}$, and $i: J^1_n V \longrightarrow \tilde{V}$ given by (3.3) and (3.4), respectively, for $p = n$.

Then, for a fixed $u \in V$, let $t: FM \longrightarrow V$ of type (ρ, V) define a G_u-structure P_{G_u} on M, and denote by F the associated tensor field on M of type $(1, s)$ or $(0, s)$, a tensor field which will be said to define P_{G_u}. Then, combining Theorem 3.2.9 with the results stated for the complete lift of tensor fields, we can assert that the complete lift F^C of F to FM defines $P_{G_{\tilde{u}}|FM}$, and conversely.

We observe that this generalizes the cases considered in 2.4 for tensor fields of types $(1,1)$ and $(0,2)$.

Next, let be $V = \mathbf{R}$ and $\rho: Gl(n,\mathbf{R}) \to Gl(\mathbf{R})$ given by

$$\rho(g)\xi = (\det(g))\xi \quad , \; g \in Gl(n,\mathbf{R}) \; , \; \xi \in \mathbf{R} \; .$$

Then, if $u = 1 \in \mathbf{R}$, $G_u = Sl(n,\mathbf{R})$, the real special linear group.

An $Sl(n,\mathbf{R})$-structure P on M is determined by a differentiable function $t: FM \to V$ of type (ρ, V), or equivalently by a volume form Ω on M, the relation between Ω and t being given by

$$\Omega_x(X_1,\ldots,X_n) = \lambda(t(y))(y^{-1}X_1,\ldots,y^{-1}X_n) \; ,$$

for $X_1,\ldots,X_n \in T_xM$, $x \in M$, $y \in FM$ such that $\pi_M(y) = x$, and

$$\lambda: \mathbf{R} \longrightarrow \Lambda^n(\mathbf{R}^n)^*$$

being the canonical isomorphism given by

$$\lambda(\xi) = \xi \, e_1 \wedge \cdots \wedge e_n \quad , \quad \xi \in \mathbf{R} \; .$$

Now, consider $J^1_n\mathbf{R} \simeq \mathbf{R}^{1+n}$, $u^1 \in J^1_n\mathbf{R}$ the 1-jet at 0 of the constant map onto $u = 1$, $\tilde{V} = \mathbf{R}$, $\tilde{\rho}: Gl(n+n^2,\mathbf{R}) \to Gl(\mathbf{R})$ also defined by the determinant, and $t = \pi_{\mathbf{R}}: J^1_n\mathbf{R} \longrightarrow \mathbf{R}$ the target map. Then (3.1) is still satisfied and, for $\tilde{u} = i(u^1)$, $\tilde{G}_u = Sl(n+n^2,\mathbf{R})$.

Therefore, from Theorem 3.2.9 it follows that, given a volume form Ω on M, there exists canonically associated a volume form $\tilde{\Omega}$ on FM.

In order to compute $\tilde{\Omega}$ from Ω, we proceed as follows.

Let (U, x^i) be chart on M and assume Ω given in U by

$$\Omega = f \, dx^1 \wedge \cdots \wedge dx^n \quad ,$$

f being nowhere zero on U; then the associated function t is given by $t = f \circ \pi_M$.

If $\tilde{t}$ denotes the complete lift of t,

$$\tilde{\Omega}_{\tilde{x}}\left(\frac{\partial}{\partial x^i}, \frac{\partial}{\partial x^i_\alpha}\right) = \lambda(\tilde{t}(\tilde{y}))\left(y^{-1}\left(\frac{\partial}{\partial x^i}\right), y^{-1}\left(\frac{\partial}{\partial x^i_\alpha}\right)\right) ,$$

for any $\tilde{x} \in J^1_nU$ and $\tilde{y} \in FJ^1_nU$ such that $\pi_{J^1_nM}(\tilde{y}) = \tilde{x}$.

Thus, if $\tilde{y} = j_M(x^i, X^i_j, x^i_\alpha, X^i_{j\alpha})$ with $(x^i, X^i_j, x^i_\alpha, X^i_\alpha) \in J^1_n P_{|J^1_n U}$, P being the $Sl(n, \mathbf{R})$–structure determined by Ω, then $det(X^i_j) = 1$ and $\tilde{t}(\tilde{y}) = f(x^i)$. Therefore

$$\begin{aligned}
&\tilde{\Omega}_{\tilde{x}}\left(\frac{\partial}{\partial x^i}, \frac{\partial}{\partial x^i_\alpha}\right) \\
&\quad = f(x^i)\, e^1 \wedge \cdots \wedge e^{n+n^2}\left(\overline{X}^j_i e_j - \sum_{\beta=1}^{n} \overline{X}^h_i X^k_{h\beta} \overline{X}^j_k e_{\beta n+j} + \sum_{\beta=1}^{n} \delta^\beta_\alpha X^j_i e_{\beta n+j}\right) \\
&\quad = f(x^i)(det\,(\tilde{X}^i_j))^{n+1} = f(x^i) \ .
\end{aligned}$$

Thus, $\tilde{\Omega}$ is locally given in $J^1_n U$ by

$$\tilde{\Omega} = f\, dx^1 \wedge \cdots \wedge dx^n \wedge dx^1_1 \wedge \cdots \wedge dx^n_n \ .$$

Chapter 4

Prolongation of linear connections

Introduction

The left-invariant tangent vector fields on a Lie group G constitute a vector space $\mathfrak{g}$ isomorphic to the tangent space to G at the identity. Moreover, $\mathfrak{g}$ becomes a Lie algebra under the bracket operation for tangent fields. Intuitively, any neighbourhood of the origin in $\mathfrak{g}$ gives a linear approximation to a neighbourhood of the identity in the manifold G, and hence gives a linearized view of the neighbourhood of any point in G. So, passing to the Lie algebra of a Lie group allows us to exploit the simplicity of linear algebra in studying G and the spaces on which it acts.

Accordingly, we are interested in Lie algebra-valued differential forms and here, in particular, the $\mathfrak{g}$-valued 1-forms on a principal G-bundle because they include connections. We consider the lifting of connections, and their covariant derivative operators, to the frame bundle. It turns out that the prolongation processes preserve parallelism of tensor fields, adaptation to G-structures, geodesics, Jacobi vector fields and curvature.

4.1 Forms with values in a Lie algebra

Let $\mathfrak{g}$ be the Lie algebra of a Lie group G; in the sequel, we identify $\mathfrak{g}$ with the tangent space T_eG at e, the unit element of G.

Definition 4.1.1 *A differentiable map $\omega: \oplus_r TM \to TG$ which is a linear skew-symmetric map of $\oplus_r T_xM$ with values in T_eG for each $x \in M$ will be called a $\mathfrak{g}$-valued r-form on M.*

As in Section 3.2, the prolongation $\omega_1: \oplus_r TJ^1_pM \longrightarrow TJ^1_pG$ of ω to J^1_pM is defined by

$$\omega_1 = \alpha_G^{1,p} \circ \omega^1 \circ \simeq \circ(\oplus_r \alpha_M^{p,1}) ,$$

and it is a $J^1_p\mathfrak{g}$–valued r–form on J^1_pM. Moreover, (3.9)-(3.11) are still valid here, and a bracket product of $\mathfrak{g}$–valued forms ω, τ can be defined canonically. In fact,

$$[\omega, \tau]_1 = [\omega_1, \tau_1] . \tag{4.1}$$

Let $\rho: G \longrightarrow Aut\,(G)$ be a homomorphism of Lie groups; for each $g \in G$,

$$T\rho(g): TG \longrightarrow TG$$

induces an automorphism

$$T\rho(g): \mathfrak{g} \longrightarrow \mathfrak{g}$$

and, therefore, there is an induced representation $\rho: G \longrightarrow Aut\,(\mathfrak{g})$. Assume G acting on M on the right; a $\mathfrak{g}$–valued r–form ω on M is called of type $(\rho, \mathfrak{g})$ if, for any $g \in G$,

$$\omega \circ (\oplus_r TR_g) = T\rho(g^{-1}) \circ \omega .$$

Thus, looking back to Section 3.2, the following natural question arises: *is the prolongation ω_1 of ω a form of type $(\rho_1, J^1_p\mathfrak{g})$ for some representation $\rho_1: J^1_pG \longrightarrow Aut\,(J^1_p\mathfrak{g})$ canonically induced from ρ?*.

To answer this question we proceed as follows.

Let $Aut\,(J^1_pG)$ be the group of automorphisms of J^1_pG; then, if we adapt Section 1.4 here with the obvious changes, an injective homomorphism of groups is defined by

$$j_p: J^1_p Aut\,(G) \longrightarrow Aut\,(J^1_pG) ,$$

and thus, given

$$\rho: G \longrightarrow Aut\,(J^1_pG)$$

and the induced

$$\rho^1: J^1_pG \longrightarrow J^1_p Aut\,(G) ,$$

we set $\rho_1 = j_p \circ \rho^1$; now, we define the linear representation

$$\rho_1: J^1_pG \longrightarrow Aut\,(J^1_p\mathfrak{g})$$

by

$$\rho_1(\tilde{g}) = T\rho_1(\tilde{g}) , \quad \tilde{g} \in J^1_pG ,$$

where

$$T\rho_1(\tilde{g}): J^1_p\mathfrak{g} \longrightarrow J^1_p\mathfrak{g}$$

denotes the linear map induced by

$$\rho_1(\tilde{g}): J^1_pG \longrightarrow J^1_pG .$$

Note that there is no problem with the differentiability of j_p.

Lemma 4.1.2 *Let $\eta: \mathbf{R}^p \to G$ be differentiable, $g = j^1\eta \in J^1_pG$ and $f: \mathbf{R}^p \to TG$ such that $f(\mathbf{R}^p) \subset T_eG$. Define a map*

$$T\rho(\eta^{-1}) * f: \mathbf{R}^p \longrightarrow TG$$

as in Lemma 3.1.2. *Then:*

(i): $\alpha_G^{1,p}(j^1 f) \in T_{\tilde{e}}J^1_pG$, $\tilde{e} =$ *unit element of* J^1_pG ;

(ii): $\alpha_G^{1,p}(j^1(T\rho(\eta^{-1}) * f) = T\rho_1(\tilde{g}^{-1})(\alpha_G^{1,p}(j^1 f))$.

Theorem 4.1.3 *Let ω be a $\mathfrak{g}$-valued r-form on M of type $(\rho, \mathfrak{g})$. Then, the prolongation ω_1 of ω to J^1_pM is an r-form of type $(\rho_1, J^1_p\mathfrak{g})$.*

The proofs of both Lemma 4.1.2 and Theorem 4.1.3 are similar to those of Lemma 3.1.2 and Theorem 3.2.5, respectively.

Let $\rho: G \longrightarrow Aut\,(G)$ be given by $\rho(g) = R_{g^{-1}} \circ L_g$, $g \in G$; then, the induced representation $\rho: G \longrightarrow Aut\,(\mathfrak{g})$ is the adjoint representation, i.e. $T\rho(g) = ad\ g$. Therefore, for any $g = j^1\eta$, $\tilde{\mu} = j^1\mu \in J^1_p\mathfrak{g}$,

$$\rho_1(\tilde{g})(\tilde{\mu}) = j_p(j^1(\rho \circ \eta))(\tilde{\mu}) = j^1((\rho \circ \eta) * \mu) \ ,$$

where

$$\begin{aligned} ((\rho \circ \eta) * \mu)(t) &= \rho(\eta(t))(\mu(t)) \\ &= \eta(t) \circ \mu(t) \circ \eta(t)^{-1} \\ &= (\eta \cdot \mu \cdot \eta^{-1})(t) \ , \ t \in \mathbf{R}^p \ ; \end{aligned}$$

therefore $j^1((\rho \circ \eta) * \mu) = \tilde{g} \cdot \tilde{\mu} \cdot \tilde{g}^{-1}$ and, hence, $\rho_1(\tilde{g}) = R_{\tilde{g}^{-1}} \circ L_{\tilde{g}}$. Thus the following lemma is proved:

Lemma 4.1.4 *Let $ad: G \longrightarrow Aut\,(\mathfrak{g})$ be the adjoint representation. Then*

$$(ad)_1: J^1_pG \longrightarrow Aut\,(J^1_p\mathfrak{g})$$

is the adjoint representation of $J^1_p\mathfrak{g}$.

Corollary 4.1.5 *Let ω be a $\mathfrak{g}$-valued r-form on M of type $(ad, \mathfrak{g})$. Then, the prolongation ω_1 of ω to J^1_pM is of type $(ad, J^1_p\mathfrak{g})$.*

4.2 Prolongation of connections to principal fibre bundles

Let $P(M, \pi, G)$ be a principal fibre bundle, and let $\omega: TP \longrightarrow TG$ be a $\mathfrak{g}$-valued 1-form on P. Denote by $L_x: G \longrightarrow P$ the differentiable map given by $L_x(g) = xg$, for each $x \in P$. We are interested in ω satisfying:

$$\omega(TL_x(v)) = TL_{g^{-1}}(v) \ , \ v \in T_gG \ , \ g \in G \ . \tag{4.2}$$

Lemma 4.2.1 *If ω satisfies* (4.2), *then its prolongation ω_1 does too, i.e.*

$$\omega_1(TL_{\tilde{x}}(\tilde{v})) = TL_{\tilde{g}^{-1}}(\tilde{v}) \ , \tag{4.3}$$

for every $\tilde{x} \in J^1_pP$, $\tilde{g} \in J^1_pG$ and $\tilde{v} \in T_{\tilde{g}}J^1_pG$.

Proof. Let $\phi: \mathbf{R} \longrightarrow J^1_pG$ and $\psi: \mathbf{R} \times \mathbf{R}^p \longrightarrow G$ be such that $\tilde{v} = \dot{\phi}$, $\tilde{g} = \phi(0) = j^1\psi_0$ and $\phi(t) = j^1\psi_t$ for small t, where $\psi_t(u) = \psi^u(t) = \psi(t,u)$, $t \in \mathbf{R}$, $u \in \mathbf{R}^p$. Define $\psi': \mathbf{R} \times \mathbf{R}^p \longrightarrow G$ by $\psi'(t,u) = \psi(0,u)^{-1}\psi(t,u)$; then $(L_{\tilde{g}^{-1}} \circ \phi)(t) = j^1\psi'_t$ and, if we set $\Psi': \mathbf{R}^p \longrightarrow TG$ given by $\Psi'(u) = \dot{\psi}'^u$, then

$$\alpha^{p,1}_G(TL_{\tilde{g}^{-1}}(\tilde{v})) = j^1\Psi' \ . \tag{4.4}$$

On the other hand, let $\eta: \mathbf{R}^p \longrightarrow P$ be such that $\tilde{x} = j^1\eta$, and let us define $\eta': \mathbf{R} \times \mathbf{R}^p \to P$ by $\eta'(t,u) = \eta(u)\psi(t,u)$; then $(L_x \circ \phi)(t) = j^1\eta'_t$ for small t, and if we set $Y: \mathbf{R}^p \longrightarrow TP$ given by $Y(u) = \dot{\eta}'^u$, then $\alpha^{p,1}_P(TL_{\tilde{x}}(\tilde{v})) = j^1Y$. Therefore

$$(\omega^1 \circ \alpha^{p,1}_P)(TL_{\tilde{x}}(\tilde{v})) = j^1(\omega \circ Y) \ . \tag{4.5}$$

Now, since $\eta'^u(t) = \eta(u)\psi^u(t)$, $\dot{\eta}'^u = TL_{\eta(u)}(\dot{\psi}^u)$, and hence, in view of (4.2),

$$(\omega \circ Y)(u) = \omega(TL_{\eta(u)}(\dot{\psi}^u)) = TL_{\psi^u(0)}(\dot{\psi}^u) \ .$$

On the other hand $\psi'^u(t) = \psi^u(0)^{-1}\psi^u(t)$; therefore $\Psi'(u) = (\omega \circ Y)(u)$ and hence $\Psi' = \omega \circ Y$. Finally, by (4.4) and (4.5),

$$(\omega^1 \circ \alpha^{p,1}_P)(TL_{\tilde{x}}(\tilde{v})) = \alpha^{p,1}_G(TL_{\tilde{g}^{-1}}(\tilde{v})) \ ,$$

and the lemma is proved.□

Definition 4.2.2 *Let Γ be a connection in $P(M,\pi,G)$, and let ω be the connection form of Γ. Following Kobayashi* [44], *ω can be considered as a differential 1-form on P of type $(ad, \mathfrak{g})$ which satisfies* (4.2).

If, as before, ω_1 denotes the prolongation of ω to J^1_pP, trom Corollary 4.1.5 and (4.1), we deduce

Theorem 4.2.3 *ω_1 defines a connection Γ_1 on $J^1_pP(J^1_pM, \pi^1, J^1_pG)$, and it will be called the prolongation of connection Γ on P.*

Remark 4.2.4 Note that, for $p = 1$, ω_1 coincides with the so called tangent connection of ω introduced by Kobayashi [44], also obtained by Morimoto [69].

Local expressions

For later use, let us describe the local expressions of ω and ω_1.

Let (U, x^i) and (U', y^a) be local charts on P and G respectively, $1 \leq i \leq dim\ P$, $1 \leq a \leq dim\ G$, and $(TU, x^i, \dot{x}^i)$, $(TU', y^a, \dot{y}^a)$ the induced charts on TP and TG respectively. With respect to these charts, ω is given by

$$\omega: \ y^a = \omega^a(x^i, \dot{x}^i) = y^a(e) \ , \quad \dot{y}^a = \dot{\omega}(x^i, \dot{x}^i) \ ,$$

and therefore, for every i and a,

$$\frac{\partial \omega^a}{\partial x^i} = \frac{\partial \omega^a}{\partial \dot{x}^i} = 0 \ .$$

Now, if $(x^i, x^i_\alpha; \dot{x}^i, \dot{x}^i_\alpha)$ and $(y^a, y^a_\alpha; \dot{y}^a, \dot{y}^a_\alpha)$ are the coordinate functions on TJ^1_pU and TJ^1_pU' respectively, and taking into account the local expressions of $\alpha^{p,1}_P$, $\alpha^{1,p}_G$ and ω^1 (see 1.1 and (1.5)), a straightforward computation leads to the following local expression for ω_1:

$$\omega_1 : \begin{cases} y^a = y^a(e) \ , \qquad y^a_\alpha = 0; \\ \dot{y}^a = \dot{\omega}^a(x^i; \dot{x}^i) \ ; \\ \dot{y}^a_\alpha = \dfrac{\partial \dot{\omega}^a}{\partial x^k}(x^i; \dot{x}^i)\, x^k_\alpha + \dfrac{\partial \dot{\omega}^a}{\partial \dot{x}^k}(x^i; \dot{x}^i)\, \dot{x}^k_\alpha \ . \end{cases}$$

Theorem 4.2.5 *Let Ω be the curvature form of connection Γ on P. Then, the prolongation Ω_1 of Ω is the curvature form of the prolongation Γ_1 of Γ. Therefore, Γ is flat if and only if Γ_1 is flat.*

Proof. By virtue of (3.10) and (4.1), and the structure equation of Γ, $\Omega = d\omega + (1/2)[\omega, \omega]$, we have

$$\Omega_1 = (d\omega)_1 + (1/2)[\omega, \omega]_1 = d\omega_1 + (1/2)[\omega_1, \omega_1] \ ,$$

which proves the theorem.□

Covariant differentiation operators

Next, and again with a view to further applications, we shall search for the relation between the operators of exterior covariant differentiation D and D_1 with respect to Γ and Γ_1 respectively.

So, let V be again a vector space, $\rho: G \longrightarrow Gl(V)$ a linear representation and τ a V-valued r-form on P of type (ρ, V); recall that τ is called *tensorial* if $\tau(X_1, \ldots, X_r) = 0$ whenever at least one of the tangent vectors X_i of P is vertical.

Proposition 4.2.6 *The prolongation τ_1 of a tensorial form τ on P of type (ρ, V) is also a tensorial form on $J^1_p P$.*

Proof. Routine, taking into account the local expressions of τ and τ_1 (see Section 3.2).□

Definition 4.2.7 *Let X be a vector field on M and Γ a connection in the principal fibre bundle $P(M, \pi, G)$. The horizontal lift X^H of X to P with respect to Γ is the unique vector field on P that is both horizontal and projectable, and with projection X.*

Let Γ be a connection in $P(M, \pi, G)$, Γ_1 the prolongation of Γ to the principal bundle $J^1_p P(J^1_p M, \pi^1, J^1_p G)$, and for a vector field X let us denote by X^H its horizontal lift with respect to Γ or Γ_1. Then:

Proposition 4.2.8 *For any vector field X on M,*

$$(4.6) \qquad \begin{aligned} (X^C)^H &= (X^H)^C , \\ (X^{(\alpha)})^H &= (X^H)^{(\alpha)} , \quad 1 \leq \alpha \leq p . \end{aligned}$$

Proof. From (3.8),

$$\omega_1((X^H)^C) = (\omega(X^H))^1 = 0 , \quad \omega_1((X^H)^{(\alpha)}) = (\omega(X^H))^{(\alpha)} = 0 ,$$

and therefore $(X^H)^C$, $(X^H)^{(\alpha)}$ are all horizontal with respect to Γ_1. Hence, it suffices to check that, at any point $\tilde{x} \in J^1_p P$,

$$T\pi^1((X^H)^C_{\tilde{x}}) = (X^C)_{\pi^1(\tilde{x})} , \quad T\pi^1((X^H)^{(\alpha)}_{\tilde{x}}) = (X^{(\alpha)})_{\pi^1(\tilde{x})} ,$$

which can be done using local expressions.□

Proposition 4.2.9 *For any V-valued r-form τ on P of type (ρ, V)*

$$(4.7) \qquad D_1\tau_1 = (D\tau)_1 .$$

Proof. Since $(D\tau)_1$ is tensorial, it suffices to check the identity by applying both sides to horizontal arguments of the form $(X^C)^H$ and $(X^C)^{(\alpha)}$ for arbitrary vector fields X on M; the result follows directly from the definition of D and D_1 taking into account (3.9), (3.10) and (4.6).□

4.3 Complete lift to FM of linear connections

The results in Section 4.2 can be specialized to the principal fibre bundle of linear frames on M.

Let $FM(M, \pi_M, Gl(n, \mathbf{R}))$ be the bundle of linear frames on M and Γ a linear connection on M. Then Γ_1, prolongation of Γ to FM, is a connection on the principal bundle $J^1_n FM(J^1_n M, \pi^1, J^1_n Gl(n, \mathbf{R}))$; it induces a linear connection $\tilde{\Gamma}$ on FM as follows: let

$$j_M : J^1_n FM \to F J^1_n M$$

be the canonical imbedding (given in Theorem 2.1.2), and let $\tilde{\Gamma}$ be the unique linear connection on $J^1_n M$ whose connection form ω satisfies

$$j^*_M \tilde{\omega} = T j_n \circ \omega_1 \ .$$

Then, Γ^C is the restriction of $\tilde{\Gamma}$ to the open submanifold $FM \subset J^1_n M$.

Definition 4.3.1 *The linear connection Γ^C on FM is called the complete lift of Γ to FM.*

Remark that if $\tilde{\Omega}$ and Ω are the curvature forms of $\tilde{\omega}$ and ω, respectively, then $j^*_M \tilde{\Omega} = T j_n \circ \Omega_1$ and hence, since the curvature form Ω^C of Γ^C is given by $\Omega^C = \tilde{\Omega}_{|FM}$, we deduce:

Corollary 4.3.2 *The complete lift Γ^C of Γ to FM is flat if and only if Γ is flat.*

Let θ_M and $\theta_{J^1_n M}$ be the canonical 1-forms of FM and $F J^1_n M$ respectively.

Lemma 4.3.3 $j^*_M \theta_{J^1_n M} = (\theta_M)_1$.

Proof. Straightforward computations using local expressions.□

Let $\Theta = D\theta_M$ be the torsion form of Γ; then

$$\Theta_1 = (D\theta_M)_1 = D_1(\theta_M)_1 = D_1(j^*_M \theta_{J^1_n M}) = j_M(\widetilde{D}\theta_{J^1_n M}) = j^*_M \tilde{\Theta} \ ,$$

where $\widetilde{D}$ denotes the exterior covariant differentiation with respect to $\tilde{\Gamma}$, and $\tilde{\Theta}$ is the torsion form of $\tilde{\omega}$. Restricting once more to FM, we get:

Theorem 4.3.4 *Let $\tilde{\Theta}$, Θ be the torsion forms of $\tilde{\omega}$ and ω respectively. Then $j^*_M \tilde{\Theta} = \Theta_1$; hence Γ is torsion free if and only if $\tilde{\Gamma}$ is also torsion free.*

Next we shall compute the local components $\widetilde{\Gamma}^A_{BC}$ of the complete lift Γ^C of Γ to FM, $1 \leq A, B, C \leq n + n^2$.

Let $\omega: TFM \longrightarrow TGl(n, \mathbf{R})$ be the connection form of Γ, (U, x^h) a chart on M, (x^h, X^h_k) the induced coordinate functions on FM (we adopt this notation to make the computations easier), (y^i_j) the canonical coordinates on $Gl(n, \mathbf{R})$, and $(x^h, X^h_k; \dot{x}^h, \dot{X}^h_k)$, $(y^i_j; \dot{y}^i_j)$ the coordinate functions induced in TFU and $TGl(n, \mathbf{R})$, respectively. Then, ω will be locally expressed by:

$$\omega : \begin{cases} y^i_j = \omega^i_j(x^h, X^h_k; \dot{x}^h, \dot{X}^h_k) = \delta^i_j \,, \\ \dot{y}^i_j = \dot{\omega}^i_j(x^h, X^h_k; \dot{x}^h, \dot{X}^h_k) \,; \end{cases}$$

thus, if $\{e^i_j\}$ denotes the natural basis of $gl(n, \mathbf{R}) \equiv T_e Gl(n, \mathbf{R})$, we can write

$$\omega(x^h, X^h_k; \dot{x}^h, \dot{X}^h_k) = \dot{\omega}^i_j(x^h, X^h_k; \dot{x}^h, \dot{X}^h_k)\, e^j_i \in T_e Gl(n, \mathbf{R}) \,.$$

Let $\sigma: U \longrightarrow FM$ be the canonical section of FM over U, that is $\sigma(x) = (x^i, \delta^i_j)$ for every $x = (x^i) \in U$, and put $\omega_U = \sigma^*\omega$; then ω_U determines the local components Γ^i_{jk} of Γ over U by the equation

$$\omega_U = (\Gamma^i_{jk}\, dx^j)\, e^k_i \,,$$

and using [47,Prop. 7.3], one easily deduces:

$$\dot{\omega}^i_j(x^h, X^h_k; \dot{x}^h, \dot{X}^h_k) = Y^i_k \Gamma^k_{hl} X^l_j \dot{x}^h + Y^i_h \dot{X}^h_j$$

where $(Y^i_k) = (X^i_k)^{-1}$. Hence, at the point $q = (x^h, X^h_k; \dot{x}^h, \dot{X}^h_k)$, we have

$$\begin{aligned}
\frac{\partial \dot{\omega}^i_j}{\partial x^k}(q) &= Y^i_j(\partial_k \Gamma^l_{hm}) X^m_j \dot{x}^h \,, \\
\frac{\partial \dot{\omega}^i_j}{\partial X^h_k}(q) &= -Y^i_h X^k_l \Gamma^l_{mr} X^r_j \dot{x}^m + Y^i_l \Gamma^l_{mh} \dot{x}^m \delta^j_k - Y^i_h X^k_m \dot{X}^m_j, \\
\frac{\partial \dot{\omega}^i_j}{\partial \dot{x}^k}(q) &= Y^i_l \Gamma^l_{kh} X^h_j, \\
\frac{\partial \dot{\omega}^i_j}{\partial \dot{X}^h_k}(q) &= Y^i_h \delta^j_k.
\end{aligned}$$

Now, if $\sigma^1: J^1_n U \longrightarrow J^1_n FM$ is the section induced by σ, then $\tilde{\sigma} = j_M \circ \sigma^1$ is the natural section of $FJ^1_n M$ over $J^1_n U$. Thus, if $\widetilde{\Gamma}^A_{BC}$ denotes the local components of $\widetilde{\Gamma}$ on $J^1_n M$ and $\tilde{\omega}$ denotes its connection form, then, with respect to the induced coordinates (x^i, x^i_α) on $J^1_n U$ and the canonical basis $\{E^A_B\}$ of $gl(n+n^2, \mathbf{R}) \equiv T_e Gl(n+n^2, \mathbf{R})$, one has

$$\tilde{\omega}\left(\left(\frac{\partial}{\partial x^j}\right)_{\tilde{u}}\right) = \tilde{\Gamma}^A_{jB} E^B_A \,, \quad \tilde{\omega}\left(\left(\frac{\partial}{\partial x^j_\alpha}\right)_{\tilde{u}}\right) = \tilde{\Gamma}^A_{j_\alpha B} E^B_A \,.$$

at every point $\tilde{u} = \sigma(u)$ with $u \in J^1_n U$. On the other hand, setting $u_1 = \sigma^1(u)$,

$$(Tj_M)\left(\left(\frac{\partial}{\partial x^j}\right)_{u_1}\right) = \left(\frac{\partial}{\partial x^j}\right)_{\tilde{u}} \,, \quad (Tj_M)\left(\left(\frac{\partial}{\partial x^j_\alpha}\right)_{u_1}\right) = \left(\frac{\partial}{\partial x^j_\alpha}\right)_{\tilde{u}} \,,$$

hence

$$\tilde{\omega}\left(\left(\frac{\partial}{\partial x^j}\right)_{\tilde{u}}\right) = j_n\left(\omega_1\left(\left(\frac{\partial}{\partial x^j}\right)_{u_1}\right)\right) \,,$$

$$\tilde{\omega}\left(\left(\frac{\partial}{\partial x^j_\alpha}\right)_{\tilde{u}}\right) = j_n\left(\omega_1\left(\left(\frac{\partial}{\partial x^j}\right)_{u_1}\right)\right) \,,$$

Therefore, if $u = (x^i, x^i_\alpha)$, it follows

$$\begin{aligned}
\omega_1\left(\left(\frac{\partial}{\partial x^j}\right)_{u_1}\right) &= \omega_1(x^i, I, x^i_\alpha, 0; \delta^i_j, 0, 0, 0) \\
&= (I, 0; \dot{\omega}^h_i(x^i, I; \delta^i_j, 0), \frac{\partial \dot{\omega}^h_i}{\partial x^k} x^k_\alpha) \\
&= (I, 0; \Gamma^h_{ji}, x^k_\alpha(\partial_k \Gamma^h_{ji})) \,,
\end{aligned}$$

$$\begin{aligned}
\omega_1\left(\left(\frac{\partial}{\partial x^j_\gamma}\right)_{u_1}\right) &= \omega_1(x^i, I, x^i_\alpha, 0; 0, 0, \delta^\alpha_\gamma \delta^i_j, 0) \\
&= (I, 0; \dot{\omega}^h_i(x^i, I; 0, 0), \frac{\partial \dot{\omega}^h_i}{\partial \dot{x}^k} \delta^k_j \delta^\alpha_\gamma) \\
&= (I, 0; 0, \delta^\alpha_\gamma \Gamma^h_{ji})) \,,
\end{aligned}$$

where I and 0 denote, respectively, the unit matrix and the zero matrix. Hence

$$\begin{aligned}
\tilde{\omega}\left(\left(\frac{\partial}{\partial x^j}\right)_{\tilde{u}}\right) &= \Gamma^h_{ji} E^i_h + x^k_\alpha(\partial_k \Gamma^h_{ji}) E^i_{h_\alpha} + \delta^\alpha_\beta \Gamma^h_{ji} E^{i_\beta}_{h_\alpha} \,, \\
\tilde{\omega}\left(\left(\frac{\partial}{\partial x^j_\gamma}\right)_{\tilde{u}}\right) &= \delta^\alpha_\gamma \Gamma^h_{ji} E^i_{h_\alpha} \,,
\end{aligned}$$

where again $i_\alpha = \alpha n + i$; if we now restrict to the coordinate neighbourhood FU, we get the local components of Γ^C in FM:

$$
(4.8) \qquad \begin{cases} \widetilde{\Gamma}^h_{ji} = \Gamma^h_{ji}, \widetilde{\Gamma}^h_{ji_\beta} = \widetilde{\Gamma}^h_{j_\gamma i} = \widetilde{\Gamma}^h_{j_\gamma i_\beta} = \widetilde{\Gamma}^{h_\alpha}_{j_\gamma i_\beta} = 0 \,, \\ \widetilde{\Gamma}^{h_\alpha}_{ji} = x^k_\alpha(\partial_k \Gamma^h_{ji}), \widetilde{\Gamma}^{h_\alpha}_{ji_\beta} = \delta^\alpha_\beta \Gamma^h_{ji} \,, \; \widetilde{\Gamma}^{h_\alpha}_{j_\beta i} = \delta^\alpha_\beta \Gamma^h_{ji} \,. \end{cases}
$$

Comparing with the definition by Mok [65] of the complete lift of a connection to the fibre bundle of linear frames FM of M, one easily checks that that lift coincides with the one we have defined here. In fact:

Theorem 4.3.5 *Let Γ be a linear connection on M, ∇ the covariant derivative associated to Γ, Γ^C the complete lift of Γ to FM and ∇^C the covariant derivative associated to Γ^C. Then Γ^C is the unique linear connection on FM satisfying the identity*

$$
(4.9) \qquad \nabla^C_{X^C} Y^C = (\nabla_X Y)^C \,,
$$

for arbitrary vector fields X, Y on M.

Proof. Direct computations using (4.8) (see also [65]).□

Combining the previous results with those in Section 3.2, the following theorem, originally proved by Mok [65], follows immediately.

Theorem 4.3.6 *Let T_∇ and R_∇ be the tensors of torsion and curvature, respectively, of a linear connection ∇ on M, and let ∇^C be the complete lift of ∇ to FM. Then the tensors of torsion T_{∇^C} and curvature R_{∇^C} of ∇^C are the complete lifts of T_∇ and R_∇ respectively, that is:*

$$
T_{\nabla^C} = (T_\nabla)^C \,, \quad R_{\nabla^C} = (R_\nabla)^C \,.
$$

Parallelism

There is a clear relationship between the parallelism of tensor fields with respect to a linear connection on M and the parallelism of the complete lift of tensors to FM with respect to the complete lift of the connection; it can be described as follows.

Firstly, bear in mind the definition of the complete lift $\tilde{t}$ to FM of a tensor $t \colon FM \longrightarrow V$ of type (ρ, V) as given in Section 3.2.

Let Γ be a linear connection on M, Γ_1 and $\widetilde{\Gamma}$ the connections induced on $J^1_n FM$ and on $FJ^1_n M$ respectively, and D, D_1 and $\widetilde{D}$ their respective exterior covariant differentials.

Proposition 4.3.7 *Let $\tilde{t}$ be the complete lift of a tensor $t\colon FM \to V$ of type (ρ, V). Then*

$$j_M^*(\widetilde{D\tilde{t}}) = i \circ (Dt)_1 .$$

Proof. (3.13) implies $\tilde{t} \circ j_M = i \circ t^1$, and hence $j_M^*(d\tilde{t}) = i \circ (dt^1)$; then, since j_M preserves horizontal vectors, $j_M(\widetilde{D\tilde{t}}) = i \circ (Dt^1)$, and the result follows from (4.7).□

Corollary 4.3.8 *$Dt \equiv 0$ implies $\widetilde{D\tilde{t}} \equiv 0$.*

Proof. Obviously $Dt \equiv 0$ implies $j_M^*(\widetilde{D\tilde{t}}) \equiv 0$; since $\widetilde{D\tilde{t}}$ is of type $(\tilde{\rho}, \widetilde{V})$, it follows $\widetilde{D\tilde{t}} \equiv 0$.□

Although the converse of this corollary is not true in general, there are two interesting particular cases where it is true.

Theorem 4.3.9 *Let Γ be a linear connection and F a tensor field of type $(1, s)$ on M, and Γ^C and F^C their complete lifts to FM. Then F is parallel with respect to Γ if and only if F^C is parallel with respect to Γ^C.*

Proof. Let $t\colon FM \longrightarrow \otimes_s^1\mathbf{R}^n$ be the tensor associated to F; it is well known that F is parallel with respect to Γ if and only if $Dt \equiv 0$. Therefore Corollary 4.3.8 implies that if F is parallel with respect to Γ then F^C is parallel with respect to Γ^C.

Conversely, note that in this case $i\colon J_n^1V \longrightarrow \widetilde{V}$ is given by (3.3) and hence it is injective, which implies, combined with (4.7) and Theorem 4.2.5, that the converse of Corollary 4.3.8 is also valid here and, consequently, if F^C is parallel with respect to Γ^C then F is parallel with respect to Γ.□

Theorem 4.3.10 *Let Γ be a linear connection and F a tensor field on M of type $(0, s)$, and let Γ^C and F^C be their complete lifts to FM. Then F is parallel with respect to Γ if and only if F^C is parallel with respect to Γ^C.*

Proof. It is similar to that of Theorem 4.3.9, because the converse of Corollary 4.3.8 is still valid here. In fact, $\widetilde{D\tilde{t}} \equiv 0$ implies that $i(Dt)_1(X^C)^H = 0$ and $i(Dt)_1(X^{(\alpha)})^H = 0$ for every vector field X on M, and hence, in view of (3.8) and Definition 4.2.2, it follows that $i((Dt)X^H)^1 = 0$ and $i((Dt)X^H)^{(\alpha)} = 0$, or equivalently $i(X^Ht)^1 = 0$ and $i(X^Ht)^{(\alpha)} = 0$; finally (3.4) implies $X^Ht = 0$, and therefore $Dt \equiv 0$.□

Remark 4.3.11 Note that Theorems 4.3.9 and 4.3.10 can be alternatively obtained using the identity $\nabla^C F^C = (\nabla F)^C$, whose demonstration can be easily achieved by a straightforward computation using (3.18), (3.20) and (3.16).

4.4 Complete lift of connections adapted to G–structures

Let G be a Lie subgroup of $Gl(n, \mathbf{R})$, $P(M, \pi, G)$ a G-structure on M and Γ a linear connection on M.

Definition 4.4.1 *The linear connection Γ on M is called adapted to the G-structure P, or a G-connection relative to P, if Γ is reducible to a connection in $P(M, \pi, G)$.*

Lemma 4.4.2 *Let $P(M, \pi, G)$ be a reduced fibre bundle of the principal bundle $P'(M, \pi, G')$, and let ω' be a connection on P' reducible to the connection ω in P. Then $J^1_pP(J^1_pM, \pi^1, J^1_pG)$ is a reduced subbundle of $J^1_pP'(J^1_pM, \pi^1, J^1_pG')$, and the prolongation ω'_1 of ω' to J^1_pP' is reducible to the prolongation ω_1 of ω to J^1_pP.*

Proof. Let $f: P \longrightarrow P'$ be the injective homomorphism of principal fibre bundles realizing the reduction of G' to G, and let $f: G \longrightarrow G'$ be the corresponding homomorphism between their structure groups.

Then, a direct computation proves easily that the induced homomorphism of fibre bundles $f^1: J^1_pP \to J^1_pP'$ realizes the reduction of J^1_pG' to J^1_pG, and that the associated homomorphism of Lie groups is the induced one $f^1: J^1_pG \to J^1_pG'$.

On the other hand, that ω' is reducible to ω means that the following diagram commutes

$$\begin{array}{ccc} TP & \xrightarrow{\omega} & TG \\ {\scriptstyle Tf}\downarrow & & \downarrow{\scriptstyle Tf} \\ TP' & \xrightarrow{\omega'} & TG' \end{array}$$

Therefore, from 1.1, we obtain a new commutative diagram:

$$\begin{array}{ccccccc} TJ^1_pP & \xrightarrow{\alpha^{p,1}_P} & J^1_pTP & \xrightarrow{\omega^1} & J^1_pTG & \xrightarrow{\alpha^{1,p}_G} & TJ^1_pG \\ {\scriptstyle Tf^1}\downarrow & & \downarrow{\scriptstyle (Tf)^1} & & \downarrow{\scriptstyle (Tf)^1} & & \downarrow{\scriptstyle Tf^1} \\ TJ^1_pP' & \xrightarrow{\alpha^{p,1}_{P'}} & J^1_pTP' & \xrightarrow{(\omega')^1} & J^1_pTG' & \xrightarrow{\alpha^{1,p}_{G'}} & TJ^1_pG' \end{array}$$

from where it follows that ω'_1 is reducible to ω_1.□

Recalling the results of the preceding sections, the following theorem is immediate.

Theorem 4.4.3 *Let Γ be a linear connection on M adapted to a G-structure $P(M, \pi, G)$ on M. Then the complete lift Γ^C of Γ to FM is a linear connection adapted to the $\tilde{G}$-structure $\tilde{P}(FM, \pi, \tilde{G})$, the prolongation of P to FM.*

An alternative proof of this result can be obtained as follows.

Let U be an arbitrary coordinate neighbourhood on M and $\{X_i\}$ a local field of linear frames on U adapted to P. Then, if ∇ is the covariant differentiation associated to Γ on M and Y is an arbitrary vector field on M,

$$\nabla_Y X_i = Y^k \Lambda_{ki}^h X_h \,, \tag{4.10}$$

and the matrix $(Y^k \Lambda_{ki}^h)$ belongs to the Lie algebra $\mathfrak{g}$ of G, being $Y = Y^k X_k$; under this hypothesis, Γ is a connection adapted to the G–structure P (and conversely), and the coefficients Λ_{ki}^h in (4.10) are called the components of ∇ with respect to the adapted frame $\{X_i\}$.

Now, observe that, besides the identity (4.9), the complete lift ∇^C of ∇ to FM satisfies

$$\begin{cases} \nabla^C_{X^{(\alpha)}} Y^C = \nabla^C_{X^C} Y^{(\alpha)} = (\nabla_X Y)^{(\alpha)} \,, \\ \nabla^C_{X^{(\alpha)}} Y^{(\beta)} = 0 \,, \quad 1 \leq \alpha, \beta \leq n \,. \end{cases} \tag{4.11}$$

On the other hand, from Remark 2.3.4, we know that if $\{X_i\}$ is a field of local frames on U adapted to the G–structure P on M, then $\{X_i^C, X_i^{(\alpha)}\}$ is a field of local frames on FU adapted to the prolongation $\tilde{P}$ of P to FM.

Then, a simple computation leads to the following identities:

$$\begin{aligned} \nabla^C_{X_i^C} X_j^C &= \Lambda_{ij}^h X_h^C + \sum_{\alpha=1}^{n} \left(x_\alpha^k \frac{\partial \Lambda_{ij}^h}{\partial x^k} \right) X_h^{(\alpha)} \,, \\ \nabla^C_{X_i^C} X_j^{(\alpha)} &= \nabla^C_{X_i^{(\alpha)}} X_j^C = \Lambda_{ij}^h X_h^{(\alpha)} \,, \\ \nabla^C_{X_i^{(\alpha)}} X_j^{(\beta)} &= 0 \,, \end{aligned}$$

identities from which Theorem 4.3.10 follows immediately.

Remark 4.4.4 Note that Theorem 4.3.10 can be obtained, in its turn, as an immediate consequence of the results in 3.2 and Theorem 4.4.3.

Let us recall that if P is a G–structure on M and if ∇ is a connection adapted to P, then the linear holonomy group Ψ of ∇ is a subgroup of G. And conversely, if Ψ is the linear holonomy group of ∇ on M, then M admits a Ψ–structure P and ∇ is adapted to P. Therefore, having in mind the previous results, it follows:

Theorem 4.4.5 *Let Ψ be the linear holonomy group of a linear connection ∇ on M. Then the linear holonomy group of ∇^C is $\tilde{\Psi} = j_n(\Psi)$.*

Finally, recalling that a linear connection ∇ is flat if and only if its linear holonomy group is discrete, then one easily reobtains Corollary 4.3.2 as an immediate consequence of Theorem 4.4.5.

4.5 Geodesics of ∇^C

In this Section we shall characterize the geodesics in FM with respect to the complete lift ∇^C of a linear connection ∇ on M. The results we shall describe are due to K.P. Mok [65].

Let $\tilde{C} = \tilde{C}(t)$ be a curve in FM. We can consider, in a canonical form, $\tilde{C}(t)$ as a field of linear frames $X_\alpha(t)$ along the curve $C(t) = \pi(\tilde{C}(t))$ in M; hence, we shall write $\tilde{C}(t) = (C(t), X_\alpha(t))$.

Assume that $\tilde{C}$ is locally expressed by $x^A = x^A(t)$, i.e. $x^h = x^h(t)$, $x^h_\alpha = x^h_\alpha(t)$. Then $\tilde{C}$ is a geodesic of ∇^C if it satisfies

$$\frac{d^2x^A}{dt^2} + \tilde{\Gamma}^A_{BC}\frac{dx^B}{dt}\frac{dx^C}{dt} = 0 \ , \quad 1 \le A, B, C \le n + n^2 \ , \tag{4.12}$$

where $\tilde{\Gamma}^A_{BC}$ are the local components of ∇^C, given in (4.8).

Definition 4.5.1 *Let $C = C(t)$ be a geodesic of ∇ on FM, locally given by $x^i = x^i(t)$. A vector field X on M along C is called a Jacobi vector field if its local components $\{X^i\}$ satisfies the equations*

$$\frac{\delta^2X^h}{dt^2} + \frac{\delta}{dt}\left(T^h_{ji}X^j\frac{dx^i}{dt}\right) + R^h_{kji}\frac{dx^j}{dt}\frac{dx^i}{dt} = 0 \ , \tag{4.13}$$

where δ/dt denotes the intrinsic differentiation.

Let $\hat{\Gamma}$ be the symmetric part of Γ, that is $\hat{\Gamma}$ is the linear connection on M with local components $\hat{\Gamma}^h_{ji} = (1/2)(\Gamma^h_{ji} + \Gamma^h_{ij})$. Then, $C = C(t)$ is a geodesic of Γ on M if and only if it is a geodesic of $\hat{\Gamma}$.

We shall need the following

Lemma 4.5.2 *Let C be a geodesic of Γ on M. A vector field along C is a Jacobi vector field of Γ if and only if it is a Jacobi vector field of $\hat{\Gamma}$.*

Proof. Direct computation using the relation between the curvature tensors of Γ and $\hat{\Gamma}$.□

Let us now assume that Γ is torsion free, and let $\tilde{C} = \tilde{C}(t)$ be a curve in FM. Then, having in mind (4.8), we can express (4.13) equivalently as the equations

$$\begin{cases} \dfrac{d^2x^k}{dt^2} + \Gamma^h_{ji}\dfrac{dx^j}{dt}\dfrac{dx^i}{dt} = 0 \ , \\[2ex] \dfrac{d^2x^h_\gamma}{dt^2} + x^k_\gamma(\partial_k\Gamma^h_{ji})\dfrac{dx^j}{dt}\dfrac{dx^i}{dt} = 0 \ , \end{cases} \tag{4.14}$$

The first equation means that $C(t) = \pi(\tilde{C}(t))$ is a geodesic of Γ in M. The second equation can be transformed, using the first one, into

$$\frac{\delta^2 x_\gamma^h}{dt^2} + R_{kji}^h x_\gamma^k \frac{dt^j}{dt} \frac{dx^i}{dt} = 0 \ .$$

This means that the vector fields X_α of the field of frames defined along C are Jacobi vector fields.

Now, observe that, for any linear connection Γ on M, the symmetric part of Γ^C is just $(\hat{\Gamma})^C$. Then, by Lemma 4.5.2, we can remove the assumption of Γ being torsion free in the above discussion, and then the following theorem follows:

Theorem 4.5.3 *Let Γ be a linear connection on M and $\tilde{C}: \tilde{C}(t) = (C(t), X_\gamma(t))$ a curve in FM. Then $\tilde{C}$ is a geodesic of Γ^C in FM if and only if C is a geodesic of Γ in M and each X_γ is a Jacobi vector field along C.*

4.6 Complete lift of derivations

Let us now denote by $T(M) = \sum T_s^r(M)$ the algebra of tensor fields on M.

By a derivation of $T(M)$ we shall mean a map $D: T(M) \to T(M)$ which satisfies the following conditions:

(a) $D: T_s^r(M) \longrightarrow T_s^r(M)$;
(b) $D(S+T) = DS + DT$, $S, T \in T_s^r(M)$;
(c) $D(S \otimes T) = DS \otimes T + S \otimes DT$, $S, T \in T_s^r(M)$;
(d) D commutes with every contraction of a tensor field.

The set $\mathfrak{D}(M)$ of all derivations of $T(M)$ is an infinite dimensional Lie algebra over $\mathbf{R}$ with respect to the natural addition and multiplication, and the bracket operation defined by the commutator $[D, D'] = D \circ D' - D' \circ D$.

Two derivations D and D' of $T(M)$ coincide if they coincide on $T_0^0(M)$ and on $T_0^1(M)$, i.e. on the functions and on the vector fields on M. Every derivation D of $T(M)$ can be decomposed uniquely as follows:

$$D = \mathcal{L}_X + \iota_F \ ,$$

where $\mathcal{L}_X$ is the Lie derivative with respect to a vector field and ι_F is the derivation defined by a tensor field F of type $(1,1)$ on M. The set $\mathcal{L}(M)$ of Lie derivations $\mathcal{L}_X$ is a subalgebra of $\mathfrak{D}(M)$. On the other hand, the set $\mathfrak{E}(M)$ of derivations ι_F is an ideal of the Lie algebra $\mathfrak{D}(M)$.

Lemma 4.6.1

(a) *Let $\tilde{X}$, $\tilde{Y}$ be vector fields on FM such that*

$$\tilde{X} f^V = \tilde{Y} f^V \quad , \quad \tilde{X} f^{(\alpha)} = \tilde{Y}^{(\alpha)} \quad , \quad 1 \leq \alpha \leq n \quad ,$$

for any function f on M. Then $\tilde{X} = \tilde{Y}$.

(b) *Let $\widetilde{S}$, $\widetilde{T}$ be tensor fields of type (r,s), $s > 0$, on FM such that*

$$\widetilde{S}(X_1^C, \dots, X_s^C) = \widetilde{T}(X_1^C, \dots, X_s^C) ,$$

for arbitrary vector fields $X_1, \dots, X_s$ on M. Then $\widetilde{S} = \widetilde{T}$.

Proof. (a) follows by a simple computation in local coordinates; (b) is shown in [65].□

Proposition 4.6.2 *Two derivations D and D' of $T(FM)$ coincide if and only if:*

(a) *$Df^V = D'f^V$, $Df^{(\alpha)} = D'f^{(\alpha)}$, $1 \leq \alpha \leq n$, for any function f on M;*
(b) *$DY^C = D'Y^C$, for any vector field Y on M.*

Proof. It suffices to show that if $Df^V = 0$, $Df^{(\alpha)} = 0$ $(1 \leq \alpha \leq n)$, and $DY^C = 0$, for any function f and any vector field Y on M, then $D = 0$. Therefore, if $D = \mathcal{L}_{\tilde{X}} + \iota_{\tilde{F}}$ and

$$\begin{aligned} Df^V &= \mathcal{L}_{\tilde{X}} f^V = \tilde{X} f^V = 0 , \\ Df^{(\alpha)} &= \mathcal{L}_{\tilde{X}} F^{(\alpha)} = 0 , \quad 1 \leq \alpha \leq n , \end{aligned}$$

for every function on M, then from Lemma 4.6.1 it follows that $\tilde{X} = 0$. Thus, $D = \iota_{\tilde{F}}$ and hence

$$DY^C = \iota_{\tilde{F}} Y^C = \tilde{F} Y^C = 0$$

for every vector field Y on M. Again, from Lemma 4.6.1, we deduce $\tilde{F} = 0$.□

Let $D = \mathcal{L}_X + \iota_F$ be a derivation of $T(M)$. By definition, *the complete lift D^C of D to FM* is given by

$$D^C = \mathcal{L}_{X^C} + \iota_{F^C} .$$

Proposition 4.6.3 *For any function f and any vector field X on M,*

$$\begin{aligned} D^C f^V &= \{Df\}^V ; \\ D^C f^{(\alpha)} &= \{Df\}^{(\alpha)} , \quad 1 \leq \alpha \leq n ; \\ D^C f^C &= \{Df\}^C ; \end{aligned}$$

Proof. Direct computation.□

As a direct consequence of Propositions 4.6.2 and 4.6.3, it follows easily:

Proposition 4.6.4 *The map $D \longrightarrow D^C$ is a Lie algebra homomorphism of $\mathfrak{D}(M)$ into $\mathfrak{D}(FM)$.*

Let us consider the complete lift of a covariant differentiation. Let ∇ be a linear connection on M; then the covariant differentiation ∇_X with respect to a vector field X on M is a derivation of $T(M)$. Since $\nabla_X f = Xf$, for any function f on M, then

$$\nabla_X = \mathcal{L}_X + \iota_F$$

where F is the tensor field of type $(1,1)$ on M given by

$$FY = \nabla_X Y - [X,Y] = \widehat{\nabla}_Y X ,$$

where $\widehat{\nabla}$ denotes the opposite connection of ∇ (see Section 6.1).

Let $(\nabla_X)^C$ be the complete lift of ∇_X to FM. From Proposition 4.6.3 it follows that

$$\begin{aligned}
(\nabla_X)^C f^V &= (\nabla_X f)^V = \{Xf\}^V , \\
(\nabla_X)^C f^{(\alpha)} &= (\nabla_X f)^{(\alpha)} = \{Xf\}^{(\alpha)} , \quad 1 \leq \alpha \leq n , \\
(\nabla_X)^C Y^C &= (\nabla_X Y)^C ,
\end{aligned}$$

for any function f and any vector field X on M.

On the other hand, we can consider the complete lift ∇^C of ∇ to FM and the covariant differentiation $\nabla^C_{X^C}$ with respect to the complete lift X^C of X to FM. A direct computation shows:

$$(4.15) \qquad \begin{cases}
\nabla^C_{X^C} f^V = X^C f^V = \{Xf\}^V , \\
\nabla^C_{X^C} f^{(\alpha)} = X^C f^{(\alpha)} = \{Xf\}^{(\alpha)} , \quad 1 \leq \alpha \leq n , \\
\nabla^C_{X^C} Y^C = (\nabla_X Y)^C ,
\end{cases}$$

for any function f and any vector field Y on M.

From here, it follows easily:

Proposition 4.6.5 $(\nabla_X)^C = \nabla^C_{X^C}$ *for every vector field* X *on* M.

Chapter 5

Diagonal lifts

Introduction

Let $G \subset Gl(n, \mathbf{R})$ be a Lie subgroup. In the previous chapters, we have described a natural way to prolong G–structures on M to FM. Now, we shall describe a new but non canonical way for obtaining new prolongations that are essentially different from those considered before.

5.1 Diagonal lifts

Let $G \subset Gl(n, \mathbf{R})$ be a Lie subgroup, and denote by $G_0 = G \underline{\times} \cdots \underline{\times} G$ the diagonal product of $n+1$ copies of G.

Proposition 5.1.1 *Let G be a closed subgroup of $Gl(n, \mathbf{R})$ and $P(M, \pi, G)$ a G–structure on M. Then there exists a (non canonical) G_0–structure $\tilde{P}_0$ on FM, called the diagonal prolongation of P to FM.*

Proof. Firstly, note that G_0 is a closed subgroup of $\tilde{G} = j_n(J^1_n G)$ and that, by virtue of Theorem 1.2.1, the quotient space $\tilde{G}/G_0$ is diffeomorphic to the product $\times_n \mathfrak{g}$; therefore $\tilde{G}/G_0$ is isomorphic to a Euclidean space.

Let $\tilde{P}/G_0(FM, \pi, \tilde{G}/G_0)$ be the fibre bundle over FM with fibre $\tilde{G}/G_0$ and group $\tilde{G}$, which is associated to the principal fibre bundle $\tilde{P}$. Since the fibre $\tilde{G}/G_0$ is contractible, this bundle $\tilde{P}/G_0$ has a (non canonical) global section $\sigma: FM \longrightarrow \tilde{P}/G_0$, and hence the structure group $\tilde{G}$ of $\tilde{P}$ is reduced to G_0, which means that FM has a G_0–structure $\tilde{P}_0$.□

Let us note again that the prolongation $\tilde{P}_0$ of P to FM, obtained in Proposition 5.1.1, is non canonical since the section σ is not canonical. Nevertheless, there is a natural procedure for constructing $\tilde{P}_0$ if we simply assume the existence of a linear connection on M. Such a procedure can be described as follows.

Let Γ be a linear connection on M, Γ^i_{jk} its local components with respect to the local chart (U, x^i) in M. Recall that, if X is a vector field on M with local components $\{X^i\}$ on U, then *the horizontal lift X^H of X to FM* is the unique vector field on FM that, with respect to the induced local chart (FU, x^i, x^i_α), is expressed as

$$X^H = X^j \left(\frac{\partial}{\partial x^j} - \Gamma^h_{jk} x^k_\alpha \frac{\partial}{\partial x^h_\alpha} \right) . \tag{5.1}$$

Let G be a Lie subgroup of $Gl(n, \mathbf{R})$ and define the injective homomorphism of Lie groups $i: G \longrightarrow J^1_n G$ given by $i(a) = [a; 0]$ for every $a \in G$. Then $G_0 = (j_n \circ i)(G)$.

Now, let $P(M, \pi, G)$ be a G-structure on M, $\phi: U \longrightarrow P$ a local section and $\{X_j\}$ given by

$$X_j = \phi^i_j(x) \left(\frac{\partial}{\partial x^i} \right)_x \quad , \quad x \in U ,$$

the local field of frames associated to ϕ and, hence, adapted to P. Then, using the connection Γ, we can associate to ϕ a local section

$$\phi_\Gamma: FU \longrightarrow FFM$$

given by

$$\phi_\Gamma(X) = \{(X^H_j)_X, (X^{(\alpha)}_j)_X\} \ , \ X \in FU .$$

If $\phi': U' \longrightarrow P$ is another local section of P over a coordinate neighbourhood U', then there exists a differentiable function

$$\mathsf{a}: U \cap U' \longrightarrow G$$

such that

$$\phi' = \phi \circ \mathsf{a}$$

over $U \cap U'$. Then a direct computation from the definitions shows that $\phi'_\Gamma = \phi_\Gamma \circ \mathsf{a}_\Gamma$ over $FU \cap FU'$, where

$$\begin{aligned} \mathsf{a}_\Gamma: FU \cap FU' &\longrightarrow G_0 \\ X_x &\longmapsto (j_n \circ i)(\mathsf{a}(x)) \ . \end{aligned}$$

Thus, the following theorem is immediate.

Theorem 5.1.2 *Let $P(M, \pi, G)$ be a G-structure and Γ an arbitrary linear connection on M. Then there exists a G_0-structure $\tilde{P}_0$ on FM canonically induced by P and the connection Γ.*

Definition 5.1.3 *We call $\tilde{P}_0$ in* Theorem 5.1.2 *the diagonal lift of P with respect to* Γ.

Theorem 5.1.4 *Let P be a G–structure on M, Γ a linear connection, $\tilde{P}_0$ the diagonal prolongation of P with respect to Γ, and $f: M \longrightarrow M$ an automorphism of P. Then, $Ff: FM \longrightarrow FM$ is an automorphism of $\tilde{P}_0$ if and only if f is a Γ–transformation, i.e. f preserves the connection Γ.*

Proof. The direct implication follows from the fact that, if Ff preserves $\tilde{P}_0$ then $(Ff)_* X_j^H$ is still horizontal with respect to Γ. Conversely, if f preserves Γ then $(Ff)_* X^H = (f_* X)^H$ for every vector field X on M, and the result follows easily.□

The infinitesimal version of this theorem is the following.

Corollary 5.1.5 *Let X be a vector field on M which is an infinitesimal automorphism for P. Then X^C is an infinitesimal automorphism for $\tilde{P}_0$ if and only if X is an infinitesimal Γ–transformation.*

5.2 Applications

In order to describe two examples of diagonal prolongations of G–structures on M with respect to a linear connection, we observe that, if $\phi: U \longrightarrow P$ is a local section of P over a coordinate neighbourhood $U \subset M$, then the local field of frames adapted to $\tilde{P}_0$ induced by ϕ is just $\{X_j^H, X_j^{(\alpha)}\}$, and its dual field of coframes $\{\bar{\theta}^j, \bar{\theta}_\alpha^j\}$ is given by

$$(5.2) \qquad \bar{\theta}^j = \psi_i^j \, dx^i \quad , \quad \bar{\theta}_\alpha^j = \bar{\theta}^{\alpha n + j} = \psi_h^j \Gamma_{ik}^h x_\alpha^k \, dx^i + \psi_k^j \, dx_\alpha^k \ ,$$

where, once more, $(\psi_i^j(x)) = (\phi_i^j(x))^{-1}$, $x \in U$.

G–structures from $(1,1)$–tensors

Let $u \in End(\mathbf{R}^n)$, $\tilde{u} = j^1 u$ and $\widetilde{G_u} = j_n(J_n^1 G_u)$ as in 2.4. Then,

$$(G_u)_0 \subset \widetilde{G_u} \subset G_{\tilde{u}} \ ,$$

and therefore the diagonal prolongation $\tilde{P}_0$ with respect to the connection Γ of a G_u–structure P on M can be extended to a $G_{\tilde{u}}$–structure $\tilde{P}'$ on FM. Then, if F (respect. $\tilde{F}$) denotes the tensor field of type $(1,1)$ on M (respect. on FM) associated to P (respect. to $\tilde{P}'$), a computation similar to that we did in 2.4, but here taking into account the expressions of the adapted local field of frames and

(5.2), leads to the following local expression for $\tilde{F}$ with respect to the induced chart (FU, x^i, x^i_α):

$$
(5.3) \qquad \left\{ \begin{aligned} \tilde{F} = \; & F^i_j \frac{\partial}{\partial x^i} \otimes dx^j \\ & +(F^h_i \Gamma^i_{jk} - F^i_j \Gamma^h_{ik}) x^k_\alpha \frac{\partial}{\partial x^h_\alpha} \otimes dx^j \\ & +\delta^\beta_\alpha F^i_j \frac{\partial}{\partial x^i_\alpha} \otimes dx^j_\beta \end{aligned} \right.
$$

$\{F^i_j\}$ being the local components of F in (U, x^i).

G–structures from $(0,2)$–tensors

Let $u \in \otimes_2(\mathbf{R}^n)^*$ as in 2.4, and define

$$
(5.4) \qquad \begin{aligned} \tilde{u}\colon J^1_n\mathbf{R}^n \times J^1_n\mathbf{R}^n & \longrightarrow \mathbf{R} \\ ([x; X_\alpha], [y; Y_\alpha]) & \longmapsto u(x,y) + \sum_{\alpha=1}^{n} u(X_\alpha, Y_\alpha) . \end{aligned}
$$

From (5.4), it is easy to show that the matrix expression of $\tilde{u}$ is

$$
\tilde{u} = \begin{pmatrix} u & \dots & 0 \\ \vdots & \ddots & \vdots \\ 0 & \dots & u \end{pmatrix} .
$$

Hence Proposition 2.4.8 is still valid here. Moreover, a direct computation shows that $(G_u)_0 \subset G_{\tilde{u}}$, and hence the diagonal prolongation $\tilde{P}_0$ of a G_u–structure P on M with respect to the connection Γ can be extended to a $G_{\tilde{u}}$–structure $\tilde{P}'$ on FM.

Therefore, if G (respect. $\tilde{G}$) is the tensor field of type $(0,2)$ on M (respect. on FM) associated to P (respect. $\tilde{P}'$), then the standard procedure leads to the following local expression for $\tilde{G}$:

$$
(5.5) \qquad \tilde{G} = G_{ij}\, \eta^i \otimes \eta^j + \delta^{\alpha\beta} G_{ij}\, \eta^i_\alpha \otimes \eta^j_\beta ,
$$

where (G_{ij}) are the local components of G and

$$
\eta^i = dx^i \ , \quad \eta^i_\alpha = \Gamma^i_{hk} x^k_\alpha \, dx^h + dx^i_\alpha .
$$

General tensor fields

The lifts of tensor fields of types $(1,1)$ and $(0,2)$ obtained above can be extended to the general case of tensor fields of types $(1,s)$ or $(0,s)$, $s \geq 1$, as follows.

Let us begin by introducing some purely algebraic definitions.

Let $u \in (\mathbf{R}^n)^*$; then, there is an element $u' \in gl(n,\mathbf{R})^*$, $gl(n,\mathbf{R})^*$ being the dual of $gl(n,\mathbf{R})$, canonically associated to u and given by

$$u'(A) = \sum_{\alpha=1}^{n} u(A_\alpha) \, , \tag{5.6}$$

for any $A \in gl(n,\mathbf{R})$, where $A_\alpha = (A_1^\alpha, \ldots, A_n^\alpha) \in \mathbf{R}^n$ is the α–th row of the matrix A.

Analogously, take $u \in End\,(\mathbf{R}^n)$; then, canonically associated to u, there exists $u' \in End\,(gl(n,\mathbf{R}))$ given by

$$u'(A) = u \cdot A \, , \tag{5.7}$$

for any $A \in gl(n,\mathbf{R})$, where the dot denotes the usual product of matrices.

These two definitions of the u' associated to u can be extended in the following manner.

Let $u \in \otimes_s(\mathbf{R}^n)^*$, $s \geq 1$; then, there exists $u' \in \otimes_s gl(n,\mathbf{R})^*$ canonically associated to u, given by

$$u'(A_1, \ldots, A_s) = \sum_{\alpha=1}^{n} u((A_1)_\alpha, \ldots, (A_s)_\alpha) \, , \tag{5.8}$$

where $A_1, \ldots, A_s \in gl(n,\mathbf{R})$, and $(A_i)_\alpha \in \mathbf{R}^n$ denotes the α–th row of A_i. In particular, for $s = 2$, one has

$$u'(A,B) = \sum_{\alpha=1}^{n} u(A_\alpha, B_\alpha) \, , \quad A, B \in gl(n,\mathbf{R}) \, , \tag{5.9}$$

and it is easy to check that if u is symmetric (respectively, skew–symmetric) then u' also is symmetric (respect. skew–symmetric). Moreover, if $u = (u_{ij})$ is the matrix representation of u with respect to the natural basis of $\mathbf{R}^n$, then

$$u'(e_j^i, e_l^k) = \delta^{ik} u_{jl} \, ,$$

where $\{e_j^i\}$ is the natural basis of $gl(n,\mathbf{R})$. Therefore, if $rank\, u = r$ then $rank\, u' = rn$.

Analogously, let $u \in \mathbf{R}^n \otimes (\otimes_s(\mathbf{R}^n)^*)$, $s \geq 1$; then there exists $u' \in gl(n,\mathbf{R}) \otimes (\otimes_s gl(n,\mathbf{R})^*)$ canonically associated to u, given by

$$u'(A_1, \ldots, A_s) = \begin{pmatrix} u((A_1)_1, \ldots, (A_s)_1) \\ \cdots\cdots\cdots\cdots\cdots\cdots \\ u((A_1)_n, \ldots, (A_s)_n) \end{pmatrix} , \tag{5.10}$$

where $A_1, \ldots, A_s \in gl(n, \mathbf{R})$, that is the α-th row of $u'(A_1, \ldots, A_s)$ is given by

$$(u'(A_1, \ldots, A_s))_\alpha = u((A_1)_\alpha, \ldots, (A_s)_\alpha) \in \mathbf{R}^n ,$$

It is easy to check that this definition coincides with the previous one for $s = 1$.

Now, let us consider again the fibre bundle FM of linear frames on M, and let Γ be a linear connection on M. Let θ denote the canonical 1-form on FM and let ω be the connection form of Γ, considered as usual as a differential form on FM with values in $gl(n, \mathbf{R})$ and of adjoint type.

Let τ be an arbitrary differential 1-form on M; for each linear frame $p \in FM$, considered as an isomorphism of vector spaces

$$p\colon \mathbf{R}^n \longrightarrow T_xM \ , \quad x = \pi_M(p) \ ,$$

we put

$$u_p = \tau_x \circ p \ . \tag{5.11}$$

Definition 5.2.1 *The diagonal lift τ^D to FM of a 1-form τ on M is the differential 1-form given by*

$$(\tau^D)_p(X) = u_p(\theta_p(X)) + u_p(\omega'_p(X)) \ ,$$

for any $X \in T_pFM$, $p \in FM$, and $u'_p \in gl(n, \mathbf{R})^*$ being the element given by (5.6) *for the $u_p \in (\mathbf{R}^n)^*$ given by* (5.11).

In order to determine the local expression of τ^D, note that if (U, x^i) is a local chart on M and (FU, x^i, x^i_α) is the induced coordinate system on FM, then the horizontal distribution of Γ and the vertical distribution of FM are generated on FU, respectively, by the local vector fields

$$\begin{cases} D_j = \dfrac{\partial}{\partial x^j} - \Gamma^h_{ji} x^i_\alpha \dfrac{\partial}{\partial x^h_\alpha} = \left(\dfrac{\partial}{\partial x^j}\right)^H , \\ D^k_\beta = \dfrac{\partial}{\partial x^k_\beta} = \left(\dfrac{\partial}{\partial x^k}\right)^{(\beta)} , \end{cases} \tag{5.12}$$

and the corresponding field of dual coframes is given by

$$\begin{cases} \eta^j = dx^j \ , \\ \eta^k_\beta = \Gamma^k_{ji} x^i_\beta \, dx^j + dx^k_\beta \ . \end{cases} \tag{5.13}$$

If we now assume the 1–form τ given on U as $\tau = \tau_i dx^i$, then a direct computation from Definition 5.2.1 shows that, with respect to the local coframes (5.13), τ^D appears on FU as

$$\tau^D = \tau_i \eta^i + \sum_{\alpha=1}^{n} \tau_k \eta_\alpha^k ,$$

or, equivalently, with respect to the natural coframes, as

$$\tau^D = \sum_{\alpha=1}^{n} \left(\left(\tau_i + \Gamma_{ij}^k x_\alpha^j \tau_k \right) dx^i + \tau_i dx_\alpha^i \right) .$$

Definition 5.2.1 can be extended to arbitrary covariant tensor fields as follows. Let S be a tensor field on M of type $(0, s)$, $s \geq 1$; for each linear frame $p \in FM$, we set

$$u_p = S_x \circ (p \times \overset{s}{\cdots} \times p) , \quad x = \pi(p) . \tag{5.14}$$

Then,

Definition 5.2.2 *The diagonal lift S^D to FM of a tensor field S on M of type $(0, s)$, $s \geq 1$ is the tensor field of the same type given by*

$$(S^D)_p(X_1, \ldots, X_s) = u_p(\theta_p(X_1), \ldots, \theta_p(X_s)) + u_p'(\omega_p(X_1), \ldots, \omega_p(X_s)) ,$$

for any $X_1, \ldots, X_s \in T_pFM$, $p \in FM$, and where $u_p' \in \otimes_s gl(n, \mathbf{R})^*$ is the element given by (5.8) associated to the $u_p \in \otimes_s(\mathbf{R}^n)^*$ given by (5.14).

If $S_{j_1 \cdots j_s}$ are the components of S in U, then a direct calculation shows that

$$\left\{ \begin{array}{rcl} S^D & = & S_{j_1 \cdots j_s} \eta^{j_1} \otimes \cdots \otimes \eta^{j_s} \\ & & + \sum_{\alpha=1}^{n} \delta_\alpha^{\beta_1} \cdots \delta_\alpha^{\beta_s} S_{j_1 \cdots j_s} \eta_{\beta_1}^{j_1} \otimes \cdots \otimes \eta_{\beta_s}^{j_s} \end{array} \right. \tag{5.15}$$

is the local expression of S^D in FU with respect to the coframes given in (5.13).

Analogously, let S be a tensor field of type $(1,1)$ on M; for each linear frame $p \in FM$, set

$$u_p = p^{-1} \circ S_x \circ p , \quad x = \pi(p) . \tag{5.16}$$

Then,

Definition 5.2.3 *The diagonal lift S^D to FM of a tensor field S on M of type $(1,1)$ is the tensor field of type $(1,1)$ given by*

$$(S^D)_p(X) = B(u_p(\theta_p(X))) + \lambda(u_p'(\omega_p(X))) ,$$

for any $X \in T_pFM$, $p \in FM$. Here $u_p' \in End(gl(n, \mathbf{R}))$ is the element in (5.7) determined by $u_p \in End(\mathbf{R}^n)$ given by (5.16); also $B\xi$ denotes the standard horizontal vector field associated to $\xi \in \mathbf{R}^n$ and λA is the fundamental vector field associated to $A \in gl(n, \mathbf{R})$.

This definition can be extended as follows: let S be a tensor field on M of type $(1,s)$, $s \geq 1$; for each linear frame $p \in FM$, set

$$u_p = p^{-1} \circ S_x \circ (p \times \overset{s}{\cdots} \times p) \ , \ x = \pi(p) \ . \tag{5.17}$$

Then,

Definition 5.2.4 *The diagonal lift S^D to FM of a tensor field S on M of type $(1,s)$, $s \geq 1$ is the tensor field of the same type given by*

$$\begin{aligned}(S^D)_p(X_1,\ldots,X_s)\\ &= B(u_p(\theta_p(X_1),\ldots,\theta_p(X_s))) + \lambda(u'_p(\omega_p(X_1),\ldots,\omega_p(X_s)))\end{aligned}$$

for any $X_1,\ldots,X_s \in T_pFM$, $p \in FM$. Here $u'_p \in gl(n,\mathbf{R}) \otimes (\otimes_s gl(n,\mathbf{R})^*)$ is the element determined by (5.10) *for the $u_p \in \mathbf{R}^n \otimes (\otimes_s(\mathbf{R}^n)^*)$ given by* (5.17).

Again, if $S^h_{j_1\cdots j_s}$ are the local components of S in U, then

$$\left\{ \begin{aligned} S^D &= S^h_{j_1\cdots j_s} D_h \otimes \eta^{j_1} \otimes \cdots \otimes \eta^{j_s} \\ &\quad + \sum_{\alpha=1}^n \sum_{k=1}^n \delta^{\beta_1}_\alpha \cdots \delta^{\beta_s}_\alpha S^k_{j_1\cdots j_s} D^k_\alpha \otimes \eta^{j_1}_{\beta_1} \otimes \cdots \otimes \eta^{j_s}_{\beta_s} \end{aligned} \right. \tag{5.18}$$

is the local expression of S^D in FU with respect to the field of local frames given by (5.12) and (5.13).

Remark 5.2.5 Note that, for tensor fields of type $(0,2)$ and $(1,1)$, the local expressions (5.15) and (5.18) are reduced, respectively, to (5.5) and (5.3). This fact is not surprising. For a direct computation allows a check that, as happened with the notion of prolongation of G–structures defined by tensor fields on M (see Section 3.2), the diagonal lift to FM of a G–structure on M defined by a tensor field S (of type $(0,s)$ or $(1,s)$, $s \geq 1$) is the structure on FM determined by the diagonal lift S^D of S with respect to the same linear connection.

Chapter 6

Horizontal lifts

Introduction

Our purpose in this chapter is to introduce the horizontal lift (with respect to a linear connection Γ on M) of tensor fields on M of type $(1,s)$ or $(0,s)$, $s \geq 0$, to tensor fields of the same type on FM, always in such a way that our constructions be as similar as possible to that known for the tangent bundle TM. We find that a polynomial structure on M defined by a $(1,1)$ constant rank tensor field lifts to one on FM. We show also that a curve in FM is a geodesic with respect to the horizontal lift of a linear connection if and only if it projects onto a geodesic of M and its frame field vector fields have vanishing second covariant derivative. The Levi-Civita map sending a Riemannian metric to its unique metric connection prolongs faithfully to FM by means of the diagonal lift of the metric, and flatness is preserved and copreserved. Every FM is parallelizable, so admitting a canonical flat connection; but this flat connection is never torsion free. If M is flat, then a Kähler structure lifts to FM.

6.1 General theory

Let Γ be a linear connection on M with local components Γ^h_{ji}; denote by ∇ its covariant differentiation, by R its curvature tensor, by T its torsion tensor and by R^h_{kji} and T^h_{ji} their local components. The opposite connection $\widehat{\Gamma}$ of Γ has local components $\widehat{\Gamma}^h_{ji} = \Gamma^h_{ij}$, and its covariant differentiation will be denoted by $\widehat{\nabla}$, and by $\widehat{R}$ and $\widehat{T}$ their tensors of curvature and torsion, respectively.

Given a linear connection Γ on M, it is possible to define on FM two families of globally defined 1-forms θ^γ and ω^γ_σ whose local expressions with respect to the induced chart (FU, x^i, x^i_α) are:

$$(6.1)\qquad \begin{cases} \theta^\gamma = x^\gamma_i \, dx^i \ , \\ \omega^\rho_\sigma = x^\rho_h (\Gamma^h_{ji} x^i_\sigma \, dx^j + dx^h_\sigma) \ , \end{cases}$$

where $(x^\gamma_i) = (x^i_\gamma)^{-1}$. These $n + n^2$ global 1-forms are linearly independent at any point of FM; in fact, note that $\theta = (\theta^\gamma)$ is the canonical 1-form on FM and $\omega = (\omega^\rho_\sigma)$ is the connection form of Γ. Let E_α , E^μ_λ be the $n + n^2$ global vector fields on FM dual to θ^γ and ω^ρ_σ respectively. These vector fields generate, respectively, the horizontal distribution and the vertical distribution on FM, and their local expressions in FU are:

$$(6.2)\qquad \begin{cases} E_\alpha = x^i_\alpha \left(\dfrac{\partial}{\partial x^i} - \Gamma^j_{ik} x^k_\beta \dfrac{\partial}{\partial x^j_\beta} \right) \ , \\ E^\mu_\lambda = x^j_\lambda \dfrac{\partial}{\partial x^j_\mu} \ . \end{cases}$$

Take $A = (A^\alpha_\beta) \in gl(n, \mathbf{R})$; then $\lambda A = A^\alpha_\beta E^\beta_\alpha$ is the fundamental vector field on FM associated to A.

Let S be a tensor field on M of type $(1, s)$, $s \geq 1$, with local components $S^h_{j_1 \cdots j_s}$ on (U, x^i). Then, the n^{s+1} functions

$$S^h_{j_1 \cdots j_s} x^\alpha_h \, x^{j_1}_{\beta_1} \cdots x^{j_s}_{\beta_s}$$

are globally defined on FM and differentiable. Thus a tensor field γS on FM of type $(1, s-1)$ can be defined by setting

$$\gamma S = S^h_{j_1 \cdots j_s} x^\alpha_h x^{j_1}_{\beta_1} \cdots x^{j_s}_{\beta_s} \, E^{\beta_1}_\alpha \otimes \theta^{\beta_2} \otimes \cdots \otimes \theta^{\beta_s} \ .$$

Locally, with respect to (FU, x^i, x^i_α), γS is given by

$$(6.3)\qquad \gamma S = S^h_{j_1 \cdots j_s} x^{j_1}_\beta \, \frac{\partial}{\partial x^h_\beta} \otimes dx^{j_2} \otimes \cdots \otimes dx^{j_s} \ .$$

Observe that, in particular, if F is a tensor field of type $(1,1)$ on M, then γF is a vertical vector field on FM given by

$$(6.4)\qquad \gamma F = F^h_j x^j_\beta x^\alpha_h \, E^\beta_\alpha \ ,$$

or, equivalently, by

$$(6.5)\qquad \gamma F = F^h_j x^j_\beta \, \frac{\partial}{\partial x^h_\beta} \ ,$$

$\{F^h_j\}$ being the local components of F in U.

Similarly, for any differential 1-form τ on M, locally expressed by $\tau = \tau_i\, dx^i$, the differentiable function on FM given by

$$\gamma\tau = \sum_{\alpha=1}^{n} x^i_\alpha \tau_i \,, \tag{6.6}$$

is globally defined. In fact, if we recall (3.16), one easily checks the identity

$$f^C = \gamma(df)$$

for every differentiable function f on M. That provides an equivalent definition for the complete lift of differentiable functions on M to FM; this is, in fact, the definition adopted in [12].

Take F, a tensor field of type $(1,1)$ on M with local components $\{F_j^h\}$, and denote by $F^\circ = (F_\alpha^\beta)$ the $n \times n$ square matrix of functions given by

$$F_\alpha^\beta = F_j^h x^j_\alpha x^\beta_h \,,$$

which are globally defined on FM. For each $A = (A^\alpha_\gamma) \in gl(n, \mathbf{R})$, let us consider the product matrix $F^\circ A = (F_\alpha^\beta A^\alpha_\gamma)$; then, a vertical vector field on FM is given by

$$\lambda(F^\circ A) = F_\alpha^\beta A^\alpha_\gamma E^\gamma_\beta \,, \tag{6.7}$$

and remark that if $F = I$ is the identity tensor on M then $I^\circ A = A$ and $\lambda(I^\circ A) = \lambda A$, the fundamental vector field associated to A. Also, if $A = I$ is the unit matrix, then $\lambda(F^\circ I) = \gamma F$ as given by (6.5).

Using the definitions, the following identities can be easily checked. Take $A \in gl(n, \mathbf{R})$, f a differentiable function and F a tensor field of type $(1,1)$ on M; then, using (2.15), it follows:

$$\left\{ \begin{array}{lcl} F^C(\lambda A) & = & \lambda(F^\circ A) \,, \\ {[X^C, \lambda A]} & = & 0 \,, \\ (\lambda A)(f^C) & = & \sum_{\alpha=1}^{n} (x^i_\beta \partial_i f)\, A^\beta_\alpha \,. \end{array} \right. \tag{6.8}$$

6.2 Applications

Let Γ be a linear connection and X a vector field on M, X^C and X^H the complete and the horizontal lifts of X to FM, locally given by (1.3) and (5.1), respectively, that is

$$\begin{aligned} X^C &= X^i \frac{\partial}{\partial x^i} + x^j_\alpha \frac{\partial X^i}{\partial x^j} \frac{\partial}{\partial x^i_\alpha} \,, \\ X^H &= X^i \frac{\partial}{\partial x^i} - X^i \Gamma^h_{ij} x^j_\alpha \frac{\partial}{\partial x^h_\alpha} \,, \end{aligned}$$

Hence,

$$\begin{aligned} X^C - X^H &= \left(x^i_\alpha \frac{\partial X^h}{\partial x^i} + x^i_\alpha \Gamma^h_{ji} x^j \right) \frac{\partial}{\partial x^h_\alpha} \\ &= (x^i_\alpha \widehat{\nabla}_i X^h) \frac{\partial}{\partial x^h_\alpha} = (\widehat{\nabla}_i X^h) x^i_\alpha x^\beta_h E^\alpha_\beta , \end{aligned}$$

where the differentiable functions $(\widehat{\nabla}_i X^h) x^i_\alpha x^\beta_h$ are globally defined on FM and canonically associated to the tensor field $\widehat{\nabla} X$ of type $(1,1)$, the covariant derivative of X with respect to the opposite connection $\widehat{\Gamma}$.

Adopting a notation similar to that in TM (see [87,88]), we write

$$X^H = X^C - \widehat{\nabla}_\gamma X , \tag{6.9}$$

remarking that $\widehat{\nabla}_\gamma X = \gamma(\widehat{\nabla} X)$, in accordance with (6.3).

Identity (6.9) and the theory of horizontal lifts to the tangent bundle suggest the following generalization to the frame bundle.

Tensor fields

Let S be a tensor field on M of type (r,s) with $r = 0, 1$, and $s \geq 0$. Suppose that the components of S in (U, x^i) are $S_{j_1 \cdots j_s}$ for $r = 0$, or $S^h_{j_1 \cdots j_s}$ for $r = 1$; then, the covariant derivative ∇S of S is given by:

$$\tag{6.10} \left\{ \begin{aligned} \nabla_i S_{j_1 \cdots j_s} &= \partial_i S_{j_1 \cdots j_s} - \sum_{k=1}^n \Gamma^h_{ij_k} S_{j_1 \cdots j_{k-1} h j_{k+1} \cdots j_s} ; \\ \nabla_i S^h_{j_1 \cdots j_s} &= \partial_i S^h_{j_1 \cdots j_s} + \Gamma^h_{ik} S^k_{j_1 \cdots j_s} \\ & \quad - \sum_{k=1}^n \Gamma^l_{ij_k} S^h_{j_1 \cdots j_{k-1} l j_{k+1} \cdots j_s} . \end{aligned} \right.$$

Then, the differentiable functions given by

$$\begin{aligned} (\nabla S)_{\alpha\beta_1 \cdots \beta_s} &= (\nabla_i S_{j_1 \cdots j_s})\, x^i_\alpha x^{j_1}_{\beta_1} \cdots x^{j_s}_{\beta_s} && \text{if } r = 0 , \\ (\nabla S)^\alpha_{\beta\gamma_1 \cdots \gamma_s} &= (\nabla_i S^h_{j_1 \cdots j_s})\, x^\alpha_h x^i_\beta x^{j_1}_{\gamma_1} \cdots x^{j_s}_{\gamma_s} && \text{if } r = 1 , \end{aligned}$$

are globally defined on FM and define tensor fields of type (r,s) on FM by setting

$$\tag{6.11} \left\{ \begin{aligned} \nabla_\gamma S &= \sum_{\alpha=1}^n (\nabla S)_{\alpha\beta_1 \cdots \beta_s}\, \theta^{\beta_1} \otimes \cdots \otimes \theta^{\beta_s} && \text{if } r = 0 , \\ \nabla_\gamma S &= (\nabla S)^\alpha_{\beta\gamma_1 \cdots \gamma_s}\, E^\beta_\alpha \otimes \theta^{\gamma_1} \otimes \cdots \otimes \theta^{\gamma_s} && \text{if } r = 1 . \end{aligned} \right.$$

Using (6.1) and (6.2), the following local expression is easily obtained for $\nabla_\gamma S$:

$$\tag{6.12} \begin{cases} \nabla_\gamma S = \sum_{\alpha=1}^{n} (x_\alpha^i \nabla_i S_{j_1 \cdots j_s})\, dx^{j_1} \otimes \cdots \otimes dx^{j_s} & \text{if } r = 0\,, \\ \nabla_\gamma S = (x_\alpha^i \nabla_i S_{j_1 \cdots j_s}^h) \dfrac{\partial}{\partial x_\alpha^h} \otimes dx^{j_1} \otimes \cdots \otimes dx^{j_s} & \text{if } r = 1\,. \end{cases}$$

We remark that if τ is an s–form on M then $\nabla_\gamma \tau$ is an s–form on FM. Thus, the following definition is natural.

Definition 6.2.1 *Let S be a tensor field on M of type $(0, s)$ or $(1, s)$, $s \geq 0$. The horizontal lift S^H of S to FM is the tensor field of the same type given on FM by*

$$\tag{6.13} S^H = S^C - \widehat{\nabla}_\gamma S\,,$$

where S^C is the complete lift of S to FM given by (3.19). *Observe that for a vector field X we reobtain* (6.9).

The following proposition is proved directly from the definitions.

Proposition 6.2.2 *Let f be a differentiable function on M. Then $f^H \equiv 0$.*

Because of its intrinsic interest, as well as for its later use, we shall consider now the particular cases of tensor fields on M of types $(1,0)$ (vector fields), $(0,1)$ (differential forms) and $(1,1)$.

Let F be a tensor field on M of type $(1,1)$ with local components $\{F_j^i\}$ in (U, x^i); then a direct computation from (6.13), using (3.19) and (6.12), leads to the following local expression for F^H in (FU, x^i, x_α^i):

$$\tag{6.14} \begin{aligned} F^H = F_j^h \frac{\partial}{\partial x^h} \otimes dx^j \; &+ \; x_\alpha^k (\Gamma_{jk}^h F_i^h - \Gamma_{ik}^h F_j^i) \frac{\partial}{\partial x_\alpha^h} \otimes dx^j \\ &+ \; \delta_\alpha^\beta F_j^h \frac{\partial}{\partial x_\alpha^h} \otimes dx_\beta^j\,. \end{aligned}$$

Comparing with (5.18) and using (5.12) and (5.13), it easily follows that $F^H = F^D$, an identity which is not valid for tensor fields of type $(1, s)$ with $s > 1$. Moreover, using the local expression of F^H with respect to the local field of frames (5.12) and (5.13), i.e. the local expression (5.18) for $s = 1$, the following proposition is immediate.

Proposition 6.2.3 *Let F be a tensor field on M of type* $(1,1)$. *Then, if F has constant rank r, F^H has constant rank $(n+1)r$.*

Let τ be a differential 1–form on M, with local expression $\tau = \tau_i\, dx^i$ in (U, x^i); then, from (3.21) and (6.12), we deduce the following local expression for τ^H in (FU, x^i, x^i_α):

$$\tau^H = \sum_{\alpha=1}^{n}(x^j_\alpha \Gamma^h_{ij}\tau_h\, dx^i + \tau_i\, dx^i_\alpha) = \sum_{\alpha=1}^{n} \tau^{H_\alpha} , \tag{6.15}$$

where

$$\tau^{H_\alpha} = x^i_\alpha \Gamma^h_{ij}\tau_h\, dx^i + \tau_i\, dx^i_\alpha .$$

Now we have $\tau^H = \tau^D$; the latter was introduced in Section 5.2.

The following formulas are easily deduced from the definitions of the different lifts:

$$\begin{aligned} X^H f^V &= (Xf)^V , \quad X^H f^C = (Xf)^C - (\widehat{\nabla}_\gamma X) f^C , \\ X^H f^{(\alpha)} &= (Xf)^{(\alpha)} - (\widehat{\nabla}_\gamma X) f^{(\alpha)} , \quad 1 \le \alpha \le n , \end{aligned} \tag{6.16}$$

for any differentiable function f and any vector field X on M;

$$\left\{ \begin{aligned} \tau^H(X^H) &= \tau^{H_\alpha}(X^H) = 0 , \quad \tau^V(X^H) = \{\tau(X)\}^V , \\ \tau^H(\lambda A) &= \tau^C(\lambda A) , \quad \tau^{H_\alpha}(X^{(\beta)}) = \delta^\beta_\alpha \{\tau(X)\}^V , \\ \tau^H(X^C) &= \{\tau(X)\}^C - (\widehat{\nabla}_\gamma \tau)(X^C) = \tau^C(\widehat{\nabla}_\gamma X) , \\ \tau^C(X^H) &= \{\tau(X)\}^C - \tau^C(\widehat{\nabla}_\gamma X) = (\widehat{\nabla}_\gamma \tau)(X^C) , \end{aligned} \right. \tag{6.17}$$

for any vector field X, any differential 1–form τ and any $A \in gl(n, \mathbf{R})$;

$$\left\{ \begin{aligned} & F^H X^H = (FX)^H , \\ & F^H(\lambda A) = F^C(\lambda A) = \lambda(F^\circ A) , \\ & F^H X^C = (FX)^C - (\widehat{\nabla}_\gamma F)(X^C) \\ & \qquad = (FX)^H + F^C(\widehat{\nabla}_\gamma X) \\ & \qquad = (FX)^H + F^H(\widehat{\nabla}_\gamma X) , \\ & F^C X^H = (FX)^C - F^C(\widehat{\nabla}_\gamma X) \\ & \qquad = (FX)^H + (\widehat{\nabla}_\gamma F)(X^H) \\ & \qquad = (FX)^H + (\widehat{\nabla}_\gamma F)(X^C) , \\ & F^H(\gamma K) = \gamma(FK) , \\ & F^H(\lambda K^0 A) = \lambda((FK)^0 A) , \end{aligned} \right. \tag{6.18}$$

for any vector field X, all tensor fields F and K of type $(1,1)$ on M and any $A \in gl(n, \mathbf{R})$.

Proposition 6.2.4 *Let K and F be tensor fields of type $(1,1)$ on M, and denote by I the identity tensor field. Then,*

(a) $F^H(\gamma K) = \gamma(FK)$;

(b) $(KF)^H = K^H F^H$ *and* $I^H = I$;

(c) *if $P(t)$ is a polynomial in one variable t,* $\{P(F)\}^H = P(F^H)$.

Proof. (a) follows by computation from (6.4) (or (6.5)) and (6.3); (c) being a direct consequence of (b), it suffices to check the identities

$$(KF)^H(X^H) = K^H(F^H(X^H)) \ , \ (KF)^H(\lambda A) = K^H(F^H(\lambda A)) \ ,$$

for any vector field X on M and any $A \in gl(n, \mathbf{R})$. The first one follows from the identity $F^H X^H = (FX)^H$, and the second one using the identity $F^H(\lambda A) = \lambda(F^\circ A)$ and the equality $K^\circ F^\circ = (KF)^\circ$. Finally, $I^H = I$ because $I^\circ A = A$.□

Corollary 6.2.5 *Let F be a tensor field on M of type $(1,1)$ and of constant rank r. Then, if F defines on M a polynomial structure of rank r and structural polynomial $P(t) = 0$, its horizontal lift F^H defines on FM a polynomial structure of constant rank $(n+1)r$ and with the same structural polynomial.*

Next, we shall state some formulas for bracket products of vector fields. To do that, let us recall that the curvature tensor R of the connection Γ is given in (U, x^i) by

$$R\left(\frac{\partial}{\partial x^j}, \frac{\partial}{\partial x^i}\right)\frac{\partial}{\partial x^k} = R^h_{kji}\frac{\partial}{\partial x^h} \ ,$$

where

$$R^h_{kji} = \frac{\partial \Gamma^h_{ik}}{\partial x^j} - \frac{\partial \Gamma^h_{jk}}{\partial x^i} + \Gamma^h_{jl}\Gamma^l_{ik} - \Gamma^h_{il}\Gamma^l_{jk} \ .$$

Moreover, if X, Y are arbitrary vector fields on M, then $R(X,Y)$ is a tensor field of type $(1,1)$, and therefore we can consider the vertical vector field $\gamma R(X,Y)$ on FM defined by (6.4) or (6.5).

Then, one obtains, by direct computation, the following formulas:

$$\text{(6.19)} \quad \begin{cases} [X^H, Y^H] = [X,Y]^H - \gamma R(X,Y) \ , \\ [X^H, \lambda A] = 0 \ , \\ [X^C, Y^H] = [X,Y]^C - [X^C, \widehat{\nabla}_\gamma Y] \\ \qquad\qquad = [X,Y]^H - \gamma R(X,Y) + [\widehat{\nabla}_\gamma X, Y^H] \ , \\ [X^H, \widehat{\nabla}_\gamma Y] = \gamma(\nabla_X(\widehat{\nabla} Y)) \ , \\ [\widehat{\nabla}_\gamma X, \widehat{\nabla}_\gamma Y] = \gamma((\widehat{\nabla} Y)(\widehat{\nabla} X) - (\widehat{\nabla} X)(\widehat{\nabla} Y)) \ , \end{cases}$$

for all vector fields X, Y on M and $A \in gl(n, \mathbf{R})$.

Also, if K and F are tensor fields of type $(1,1)$ on M, we get

$$(6.20) \quad \begin{cases} [X^H, \gamma F] = \gamma(\nabla_X F) , \\ [X^H, \lambda(F^\circ A)] = \lambda((\nabla_X F)^\circ A) , \\ [\lambda(F^\circ A), \lambda B] = [\lambda A, \lambda(F^\circ B)] = \lambda(F^\circ[A, B]) , \\ [\lambda(F^\circ A), \lambda(K^\circ B)] = \lambda((KF)^\circ AB - (FK)^\circ BA) , \end{cases}$$

for any $A, B \in gl(n, \mathbf{R})$. If $K = F$, then

$$(6.21) \quad [\lambda(F^\circ A), \lambda(F^\circ B)] = \lambda(F^\circ)^2[A, B] .$$

An interesting application of formulas of this sort is the following.

Let P be a projection tensor in M. That is, P is a non-trivial tensor field of type $(1,1)$ such that $P^2 = P$. Then, there exists on M a distribution $\mathfrak{D}$ determined by $\mathfrak{D} = Im\ P$. By using Proposition 6.2.4, we find $(P^H)^2 = P^H$, which means that P^H also is a projection tensor on FM. Thus there exists on FM a distribution $\mathfrak{D}^H$ corresponding to P^H, which will be called *the horizontal lift of* $\mathfrak{D}$. If $dim\mathfrak{D} = r$ then $dim\mathfrak{D}^H = (n+1)r$, and $\mathfrak{D}^H$ is spanned by X^H and $\lambda(P^\circ A)$, X being an arbitrary vector field belonging to $\mathfrak{D}$ and $A \in gl(n, \mathbf{R})$, because

$$P^H X^H = (PX)^H \ , \ P^H(\lambda A) = \lambda(P^\circ A) ,$$

by virtue of (6.18).

If we put $Q = I - P$ and bear in mind the previous formulas, then

$$(6.22) \quad \begin{cases} Q^H[P^H(\lambda A), P^H(\lambda B)] = 0 , \\ Q^H[P^H X^H, P^H(\lambda A)] = \lambda((Q\nabla_{PX}P)^\circ A) , \\ Q^H[P^H X^H, P^H Y^H] = \{Q[PX, PY]\}^H - \gamma(QR(PX, PY)) , \end{cases}$$

for any vector fields X, Y on M and any $A, B \in gl(n, \mathbf{R})$.

As is well known, the distribution $\mathfrak{D}$ is integrable if and only if the condition $Q[PX, PY] = 0$ is satisfied for all vector fields X and Y. Therefore, taking into account (6.22), we have

Proposition 6.2.6 *Let there be given a distribution $\mathfrak{D}$ on M corresponding to the projection tensor P. The two distributions $\mathfrak{D}$ and $\mathfrak{D}^H$ are integrable at the same time if and only if the conditions*

$$QR(PX, PY) = 0 \ , \ Q\nabla_{PX}P = 0 ,$$

are satisfied for all vector fields X, Y *on* M, *or, equivalently*

$$Q^h_l R^l_{kji} P^j_a P^i_b = 0 \ , \quad Q^h_l P^k_j \nabla_k P^l_i = 0 \ ,$$

$\{P^h_j\}$ *and* $\{Q^h_j\}$ *being the components of* P *and* $Q = I - P$, *respectively, and* $i, j, k, a, b \in \{1, 2, ..., n\}$.

To obtain meaningful formulas on Lie derivatives, let us recall that the Lie derivative $\mathcal{L}_{\widetilde{X}} \widetilde{F}$ of a tensor field $\widetilde{F}$ of type $(1,1)$ with respect to a vector field $\widetilde{X}$ is defined by

$$(\mathcal{L}_{\widetilde{X}} \widetilde{F}) \widetilde{Y} = [\widetilde{X}, \widetilde{F}\widetilde{Y}] - \widetilde{F}[\widetilde{X}, \widetilde{Y}] \ .$$

Therefore, having in mind once more the previous formulas, it follows

$$\begin{aligned}
(\mathcal{L}_{X^H} F^H)(Y^H) &= \{(\mathcal{L}_X F)Y\}^H - \gamma(R(X, FY) - FR(X,Y)) \ , \\
(\mathcal{L}_{X^H} F^H)(\lambda A) &= \lambda((\nabla_X F)^\circ A)) \ ,
\end{aligned}$$

for all vector fields X, Y, $(1,1)$–tensor fields F and $A \in gl(n, \mathbf{R})$. Thus we have

Proposition 6.2.7 *Let* X *be a vector field and* F *a tensor field of type* $(1,1)$ *on* M. *Then the condition*

$$\mathcal{L}_{X^H} F^H = 0$$

is equivalent to the conditions

$$\mathcal{L}_X F = 0 \ , \ \nabla_X F = 0 \ , \ R(X, FY) - FR(X,Y) = 0$$

for all vector fields Y *on* M. *The last condition is equivalent to*

$$X^j (R^h_{kjl} F^l_i - R^l_{kji} F^h_l) = 0$$

$\{X^j\}$ *and* $\{F^h_i\}$ *being the local components of* X *and* F, *respectively.*

Similarly, for any vector field X on M and any $A, B \in gl(n, \mathbf{R})$, we find

$$(\mathcal{L}_{\lambda A} F^H)(X^H) = 0 \ , \ (\mathcal{L}_{\lambda A} F^H)(\lambda B) = 0 \ ,$$

and hence:

Proposition 6.2.8 $\mathcal{L}_{\lambda A} F^H = 0$ *for any* $A \in gl(n, \mathbf{R})$.

The Nijenhuis tensor $N_{\widetilde{F}}$ of a tensor field $\widetilde{F}$ of type $(1,1)$ is the tensor field of type $(1,2)$ given by

$$\widetilde{N}_{\widetilde{F}}(\widetilde{X},\widetilde{Y}) = [\widetilde{F}\widetilde{X},\widetilde{F}\widetilde{Y}] - \widetilde{F}[\widetilde{F}\widetilde{X},\widetilde{Y}] - \widetilde{F}[\widetilde{X},\widetilde{F}\widetilde{Y}] + \widetilde{F}^2[\widetilde{X},\widetilde{Y}] ,$$

$\widetilde{X}$ and $\widetilde{Y}$ being arbitrary vector fields. Then, a direct computation leads to the following identities:

$$\begin{aligned} N_{F^H}(\lambda A, \lambda B) &= 0 , \\ N_{F^H}(\lambda A, X^H) &= \lambda((F\nabla_X F - \nabla_{FX}F)^\circ A) , \\ N_{F^H}(X^H, Y^H) &= \{N_F(X,Y)\}^H \\ &\quad - \gamma\Big(R(FX,FY) - FR(FX,Y) \\ &\qquad -FR(X,FY) + F^2R(X,Y)\Big) , \end{aligned}$$

for all vector fields X, Y on M and $A, B \in gl(n,\mathbf{R})$. Therefore

Proposition 6.2.9 *Let F be a tensor field on M of type $(1,1)$ and F^H its horizontal lift to FM. Then the condition*

$$N_{F^H} = 0$$

is equivalent to the conditions

$$\begin{gathered} N_F = 0 , \\ F\nabla_X F - \nabla_{FX}F = 0 , \\ R(FX,FY) - FR(FX,Y) - FR(X,FY) + F^2R(X,Y) = 0 , \end{gathered}$$

for arbitrary vector fields X, Y on M. The two last conditions can be equivalently written as

$$\begin{gathered} F_i^k \nabla_k F_j^h - F_k^h \nabla_i F_j^k = 0 , \\ R_{kab}^h F_j^a F_j^b - R_{kai}^h F_j^a F_b^h - R_{kja}^h F_i^a F_b^h + R_{kji}^a F_a^b F_b^h = 0 , \end{gathered}$$

where F_i^h are the local components of F.

Linear connections

Our purpose now is to define and study the horizontal lift ∇^H of a linear connection ∇ on M to FM.

Let us begin by recalling that if ∇ is a linear connection on M, then the complete lift ∇^C of ∇ to FM is the unique linear connection on FM satisfying (4.9):

$$\nabla^C_{X^C} Y^C = (\nabla_X Y)^C$$

for all vector fields X, Y on M; moreover, ∇^C satisfies the identities:

$$\begin{aligned} &\nabla^C_{X^{(\alpha)}} Y^C = \nabla_{X^C} Y^{(\alpha)} = (\nabla_X Y)^{(\alpha)} , \\ & \qquad\qquad\qquad\qquad\qquad\qquad (1 \le \alpha, \beta \le n) \\ &\nabla^C_{X^{(\alpha)}} Y^{(\beta)} = 0 , \end{aligned}$$

and (4.8) gives the local components of ∇^C as a function of the local components of ∇.

For later use, we shall state some identities for the complete lift ∇^C of ∇. Let T and R be the torsion and the curvature tensor, respectively, of ∇. From the formulas stated before, a routine computation establishes the following identities:

$$(6.23) \qquad \begin{cases} \nabla^C_{\lambda A} \lambda B &= \lambda(AB) , \\ \nabla^C_{\lambda A} X^C &= \lambda((\nabla X)^\circ A) , \\ \nabla^C_{X^H} \lambda A &= 0 , \\ \nabla^C_{X^C} \lambda A &= \lambda((\widehat{\nabla} X)^\circ A) , \\ \nabla^C_{\lambda A} X^H &= \lambda(T(-,X)^\circ A) , \end{cases}$$

for any vector field X on M, arbitrary $A, B \in gl(n, \mathbf{R})$, and $T(-,X)$ denoting the tensor field on M of type $(1,1)$ defined by $T(-,X)(Y) = T(X,Y)$, for any vector field Y.

If F and G are tensor fields on M of type $(1,1)$, then from (6.5) and (6.17), it follows

$$(6.24) \qquad \begin{cases} \nabla^C_{\gamma F} X^C &= \gamma((\nabla X) F) , \\ \nabla^C_{X^C} \gamma F &= \gamma(\nabla_X F + F(\widehat{\nabla} X)) , \\ \nabla^C_{\gamma F} \gamma G &= \gamma(FG) , \end{cases}$$

then using $\nabla = \widehat{\nabla} - T$, and identities (6.23), from (6.24) it follows

$$(6.25) \qquad \nabla^C_{\gamma F} X^H = \gamma(T(-,X)F) , \quad \nabla^C_{X^H} \gamma F = \gamma(\nabla_X F) ,$$

Specializing (6.24) and (6.25),

$$(6.26) \qquad \begin{cases} \nabla^C_{\gamma(\nabla X)} Y^C = \gamma((\nabla Y)(\widehat{\nabla} X)) , \\ \nabla^C_{\gamma(\widehat{\nabla} X)} \gamma(\widehat{\nabla} Y) = \gamma((\widehat{\nabla} Y)(\widehat{\nabla} X)) , \\ \nabla^C_{X^C} \gamma(\widehat{\nabla} Y) = \gamma(\nabla_X(\widehat{\nabla} Y) + (\widehat{\nabla} Y)(\widehat{\nabla} X)) , \\ \nabla^C_{\gamma(\nabla X)} Y^H = \gamma(T(-,Y)(\widehat{\nabla} X)) . \end{cases}$$

Next, using again (6.3) and the previous formulas, one obtains

$$(6.27)\quad \begin{cases} \nabla^C_{X^C} Y^H = (\nabla_X Y)^H \\ \qquad\qquad + \gamma\left(R(-,X)Y - (\nabla_X T)(Y,-) + T(-,Y)(\widehat{\nabla} X)\right) , \\ \nabla^C_{X^H} Y^H = (\nabla_X Y)^H + \gamma(R(-,X)Y - (\nabla_X T)(Y,-)) , \end{cases}$$

where $R(-,X)Y$ and $(\nabla_X T)(Y,-)$ are the tensor fields of type $(1,1)$ on M defined by

$$(R(-,X)Y)(Z) = R(Z,X)Y , \quad ((\nabla_X T)(Y,-))(Z) = (\nabla_X T)(Y,Z) .$$

From (6.3), (6.7) and (6.17), it follows

$$(6.28)\quad \begin{cases} \nabla^C_{\gamma F} \lambda A = \nabla^C_{\lambda A} \gamma F = \lambda(F^\circ A) , \\ \nabla^C_{X^H} \lambda(F^\circ A) = \lambda((\nabla_X F)^\circ A) , \\ \nabla^C_{\lambda A} \lambda(F^\circ B) = \lambda(F^\circ(AB)) . \end{cases}$$

Finally, a direct computation proves the following lemma.

Lemma 6.2.10 *Let S be a tensor field on M of type (r,s), $r = 0,1$. Then*

$$\nabla^C S^C = (\nabla S)^C .$$

Definition 6.2.11 *The horizontal lift ∇^H of ∇ to FM is the linear connection defined on FM by the following conditions*

$$(6.29)\quad \begin{cases} \nabla^H_{\lambda A} \lambda B &= \lambda(AB) , \\ \nabla^H_{\lambda A} X^H &= \lambda(T(-,X)^\circ A) , \\ \nabla^H_{X^H} \lambda A &= 0 , \\ \nabla^H_{X^H} Y^H &= (\nabla_X Y)^H , \end{cases}$$

for all vector fields X, Y on M and $A, B \in gl(n, \mathbf{R})$.

Taking into account (6.23) and (6.27), we get

$$(6.30)\quad \nabla^H = \nabla^C + \widetilde{S} ,$$

where $\widetilde{S}$ is the tensor field on FM of type $(1,2)$ which satisfies the conditions

$$(6.31)\quad \begin{cases} \widetilde{S}(\lambda A, \lambda B) &= 0 , \\ \widetilde{S}(\lambda A, X^H) &= \widetilde{S}(X^H, \lambda A) = 0 , \\ \widetilde{S}(X^H, Y^H) &= -\gamma(R(-,X)Y - (\nabla_X T)(Y,-)) . \end{cases}$$

Therefore, the local components of $\widetilde{S}$ are all zero except

$$\widetilde{S}_{ji}^{h_\gamma} = -x_\gamma^k (R_{ikj}^h - \nabla_j T_{ik}^h) \ . \tag{6.32}$$

Since the components of ∇^C are given in (4.8), from (6.30) and (6.32) it follows that the horizontal lift ∇^H of ∇ has components:

$$\begin{cases} \widetilde{\Gamma}_{ji}^{h_\gamma} = x_\gamma^k(\partial_k \Gamma_{ji}^h - R_{ikj}^h + \nabla_j T_{ik}^h) \ , \\ \widetilde{\Gamma}_{ji_\alpha}^{h_\gamma} = \delta_\alpha^\gamma \Gamma_{ji}^h \ , \quad \widetilde{\Gamma}_{j_\beta i}^{h_\gamma} = \delta_\beta^\gamma \Gamma_{ji}^h \ , \\ \widetilde{\Gamma}_{ji}^h = \Gamma_{ji}^h \ , \quad \widetilde{\Gamma}_{ji_\alpha}^h = \widetilde{\Gamma}_{j_\beta i}^h = \widetilde{\Gamma}_{j_\beta i_\alpha}^h = \widetilde{\Gamma}_{j_\beta i_\alpha}^{h_\gamma} = 0 \ . \end{cases} \tag{6.33}$$

Moreover, from (6.30) and (6.31), it follows

Proposition 6.2.12 *∇^C and ∇^H are equal if and only if the tensor field W on M of type $(1,3)$ given by*

$$W(X,Y,Z) = R(Z,X)Y - (\nabla_X T)(Y,Z)$$

vanishes identically, X, Y, Z being arbitrary vector fields on M.

Corollary 6.2.13 *If ∇ has parallel torsion, i.e. $\nabla T = 0$ (resp. zero curvature, i.e. $R = 0$), then ∇^C and ∇^H are equal if and only if ∇ has zero curvature (resp. parallel torsion).*

The behaviour under parallelism with respect to ∇^H of vector fields on M is described in the following proposition.

Proposition 6.2.14 *For all $A, B \in gl(n, \mathbf{R})$ and vector fields X, Y on M,*

$$\begin{aligned} \nabla_{X^C}^H Y^C &= (\nabla_X Y)^C - \gamma(R(-,X)Y - (\nabla_X T)(Y,-)) \ , \\ \nabla_{X^C}^H Y^H &= (\nabla_X Y)^H + \gamma(T(-,Y)(\widehat{\nabla} X)) \ , \\ \nabla_{X^C}^H \lambda A &= \lambda((\widehat{\nabla} X)^\circ A) \ , \\ \nabla_{\lambda A}^H X^C &= \lambda((\nabla X)^\circ A). \end{aligned}$$

Proof. It suffices to observe that $\widetilde{S}$ vanishes when one of its arguments is a vertical vector field, and then use the previous formulas and the identity $\widetilde{S}(X^H, Y^H) = \widetilde{S}(X^C, Y^C)$.□

Similarly, one proves:

Proposition 6.2.15 *Let F be a tensor field of type $(1,1)$, X, Y vector fields on M and $A, B \in gl(n, \mathbf{R})$. Then*

$$\begin{aligned}
&\nabla^H_{\gamma F} X^H = \nabla^C_{\gamma F} X^H = \gamma(T(-,X)F) , \\
&\nabla^H_{\gamma F} \lambda A = \nabla^C_{\gamma F} \lambda A = \lambda(F^\circ A) , \\
&\nabla^H_{X^H} \lambda(F^\circ A) = \nabla^C_{X^H} \lambda(F^\circ A) = \lambda((\nabla_X F)^\circ A) , \\
&\nabla^H_{\lambda A} \lambda(F^\circ B) = \nabla^C_{\lambda A} \lambda(F^\circ B) = \lambda(F^\circ(AB)) ,
\end{aligned}$$

and hence

$$\begin{aligned}
\nabla^H_{X^H} \gamma F &= \nabla^C_{X^H} \gamma F = \gamma(\nabla_X F) , \\
\nabla^H_{\lambda A} \gamma F &= \nabla^C_{\lambda A} \gamma F = \lambda(F^\circ A) .
\end{aligned}$$

Let $\widetilde{T}$ and $\widetilde{R}$ be the torsion tensor and the curvature tensor of ∇^H, respectively; then, from (6.19) and (6.29), it follows

$$(6.34) \qquad \left\{ \begin{aligned}
\widetilde{T}(X^H, Y^H) &= \{T(X,Y)\}^H + \gamma R(X,Y) , \\
\widetilde{T}(\lambda A, X^H) &= \lambda(T(-,X)^\circ A) , \\
\widetilde{T}(\lambda A, \lambda B) &= 0 .
\end{aligned} \right.$$

On the other hand, if T^H denotes the horizontal lift of the torsion tensor T of ∇, from (6.3), (6.7) and (6.13), one easily deduces:

$$(6.35) \qquad \left\{ \begin{aligned}
T^H(X^H, Y^H) &= \{T(X,Y)\}^H , \\
T^H(\lambda A, X^H) &= \lambda(T(-,X)^\circ A) , \\
T^H(\lambda A, \lambda B) &= 0 .
\end{aligned} \right.$$

So, comparing (6.34) and (6.35), it follows:

Proposition 6.2.16 *The torsion tensor $\widetilde{T}$ of ∇^H is equal to T^H if and only if ∇ has zero curvature. Therefore, ∇^H is torsion free if and only if ∇ is locally flat, i.e $T = 0$ and $R = 0$.*

Similarly, a routine computation leads to the following formulas:

$$\begin{aligned}
\widetilde{R}(\lambda A, \lambda B)\lambda C &= \widetilde{R}(\lambda A, \lambda B) Z^H = \widetilde{R}(X^H, \lambda B)\lambda C = 0 , \\
\widetilde{R}(X^H, Y^H) Z^H &= \{R(X,Y)Z\}^H + \gamma(T(R(X,Y)-, Z)) , \\
\widetilde{R}(X^H, Y^H)\lambda C &= \lambda(R(X,Y)^\circ C) , \\
\widetilde{R}(X^H, \lambda B) Z^H &= \lambda(((\nabla_X T)(-, Z))^\circ B) ,
\end{aligned}$$

from where it follows:

Proposition 6.2.17 *∇^H has zero curvature if and only if ∇ has zero curvature and parallel torsion.*

Then, combining Corollary 6.2.13 and Proposition 6.2.17, one obtains:

Theorem 6.2.18 *If ∇^H is flat, then $\nabla^H = \nabla^C$.*

Geodesics

Next, we shall characterize the geodesics in FM with respect to the horizontal lift ∇^H of a linear connection ∇ given on M.

Let $\widetilde{C} = \widetilde{C}(t)$ be a curve in FM. We can consider $\widetilde{C}(t)$ as a field of linear frames $X_\alpha(t)$ along the curve $C(t) = \pi(\widetilde{C}(t))$ in M and, hence, we write $\widetilde{C}(t) = (C(t), X_\alpha(t))$.

Assume that $\widetilde{C}$ is locally expressed by $x^A = x^A(t)$, i.e. $x^h = x^h(t)$, $x^h_\alpha = x^h_\alpha(t)$; then $\widetilde{C}$ is a geodesic of ∇^H if it satisfies

$$\frac{d^2x^A}{dt^2} + \widetilde{\Gamma}^A_{BC}\frac{dx^B}{dt}\frac{dx^C}{dt} = 0 , \tag{6.36}$$

where $\widetilde{\Gamma}^A_{BC}$ are the local components of ∇^H, and $1 \leq A, B, C \leq n+n^2$. By means of (6.33), equations (6.36) are equivalent to

$$\begin{cases} \dfrac{d^2x^h}{dt^2} + \Gamma^h_{ji}\dfrac{dx^j}{dt}\dfrac{dx^i}{dt} = 0 , \\ \dfrac{d^2x^h_\gamma}{dt^2} + x^k_\gamma(\partial_k\Gamma^h_{ji} - R^h_{ikj} + \nabla_j T^h_{ik})\dfrac{dx^j}{dt}\dfrac{dx^i}{dt} \\ \qquad + \Gamma^h_{ji}\dfrac{dx^j_\gamma}{dt}\dfrac{dx^i}{dt} + \Gamma^h_{ji}\dfrac{dx^j}{dt}\dfrac{dx^h_\gamma}{dt} = 0 , \end{cases} \tag{6.37}$$

where Γ^h_{ji} are the local components of ∇ in M.

Note that the first n equations in (6.37) imply that C is a geodesic in M with respect to ∇. Moreover, if we assume that ∇ is torsion free, then the equations in (6.37) reduce to

$$\begin{cases} \dfrac{d^2x^h}{dt^2} + \Gamma^h_{ji}\dfrac{dx^j}{dt}\dfrac{dx^i}{dt} = 0 , \\ \dfrac{\delta^2x^h_\gamma}{dt^2} = 0 , \end{cases} \tag{6.38}$$

where

$$\frac{\delta^2x^h_\gamma}{dt^2} = \frac{d}{dt}\left(\frac{dx^h_\gamma}{dt} + \Gamma^h_{ji}\frac{dx^j}{dt}x^i_\gamma\right) + \Gamma^h_{lk}\frac{dx^l}{dt}\left(\frac{dx^k_\gamma}{dt} + \Gamma^k_{ji}\frac{dx^j}{dt}x^i_\gamma\right) ,$$

and thus, from (6.37) and (6.38), we have

Proposition 6.2.19 *Let ∇ be a torsion free linear connection on M, and let $\tilde{C} : \tilde{C}(t) = (C(t), X_\alpha(t))$ be a curve in FM. Then $\tilde{C}$ is a geodesic with respect to the horizontal lift ∇^H of ∇ if and only if its projection C down to M is a geodesic with respect to ∇ and each vector field X_α of the frame field defined by $\tilde{C}$ along C has vanishing second covariant derivative.*

Next, let $\tilde{C}$ be a curve in FM locally expressed by $x^h = x^h(t)$, $x^h_\gamma = x^h_\gamma(t)$, and let $\widetilde{X}$ be a vector field defined along $\tilde{C}$, and locally given as $\widetilde{X} = \widetilde{X}^h \partial_h + \widetilde{X}^h_\gamma \partial_{h_\gamma}$ with respect to the induced coordinates in FM. Then $\widetilde{X}$ is parallel with respect to the horizontal lift ∇^H of ∇ if it satisfies

$$\frac{d\widetilde{X}^A}{dt} + \widetilde{\Gamma}^A_{CB} \frac{dx^C}{dt} \widetilde{X}^B = 0 , \tag{6.39}$$

where $\widetilde{\Gamma}^A_{BC}$ are the components of ∇^H. From (6.33) it follows that (6.39) can be written as

$$\begin{cases} \frac{d\widetilde{X}^h}{dt} + \Gamma^h_{ji} \frac{dx^j}{dt} \widetilde{X}^i = 0 , \\ \frac{d\widetilde{X}^h_\gamma}{dt} + x^k_\gamma \left(\partial_k \Gamma^h_{ji} - R^h_{ikj} + \nabla_j T^h_{ik}\right) \frac{dx^j}{dt} \widetilde{X}^i \\ \qquad + \Gamma^h_{ji} \frac{dx^j_\gamma}{dt} \widetilde{X}^i + \Gamma^h_{ji} \frac{dx^j}{dt} \widetilde{X}^i_\gamma = 0 , \end{cases} \tag{6.40}$$

where Γ^h_{ji} are the components of ∇ in M.

When $\widetilde{X}$ satisfies the condition $\widetilde{X}^h = 0$, that is when $\widetilde{X}$ is vertical, then the equations (6.40) reduce to

$$\frac{d\widetilde{X}^h_\gamma}{dt} + \Gamma^h_{ji} \frac{dx^j}{dt} \widetilde{X}^i_\gamma = 0 .$$

Hence, if λA is the fundamental vector field associated to $A \in gl(n, \mathbf{R})$, that is $\lambda A = A^\alpha_\gamma x^h_\alpha (\partial / \partial x^h_\gamma)$, then λA is parallel along $\tilde{C}$ if and only if

$$A^\alpha_\gamma \left(\frac{dx^h_\alpha}{dt} + \Gamma^h_{ji} \frac{dx^j}{dt} x^i_\alpha \right) = 0 . \tag{6.41}$$

On the other hand, the vector field $X_\alpha = x^h_\alpha (\partial / \partial x^h)$, canonically defined in M along $C = \pi \circ \tilde{C}$, is parallel along C if and only if

$$\frac{dx^h_\alpha}{dt} + \Gamma^h_{ji} \frac{dx^j}{dt} x^i_\alpha = 0 . \tag{6.42}$$

From (6.41) and (6.42), and with the previous notations, we deduce:

Proposition 6.2.20 *Every fundamental vector field in FM is parallel along the curve $\tilde{C} : \tilde{C}(t) = (C(t), X_\alpha(t))$ if and only if each vector field X_α defined along C by $\tilde{C}$ is parallel along C.*

Let us now assume that $\widetilde{X}$ satisfies the condition of horizontality, i.e.

$$\widetilde{X}^h_\gamma = -\widetilde{X}^k \Gamma^h_{ki} x^i_\gamma ,$$

so if ∇ is assumed torsion free, then (6.40) is reduced to

$$(6.43) \qquad \begin{cases} \dfrac{d\widetilde{X}^h}{dt} + \Gamma^h_{ji} \dfrac{dx^j}{dt} \widetilde{X}^i = 0 , \\ x^l_\alpha \Gamma^h_{kl} \left(\dfrac{d\widetilde{X}^k}{dt} + \Gamma^k_{ji} \dfrac{dx^j}{dt} \widetilde{X}^i \right) = 0 , \end{cases}$$

and, obviously, any solution of the first equation in (6.43) is also a solution of the second one.

Now, we adopt the following notation: if X is a tangent vector at the initial point of $\tilde{C}$ (resp. of C), then $\tilde{C}(X)$ (resp. $C(X)$) will denote the vector field along $\tilde{C}$ (resp. along C) obtained by parallel displacement of X along $\tilde{C}$ (resp. along C) with respect to ∇^H (resp. to ∇).

Proposition 6.2.21 *Let ∇ be a torsion free linear connection on M, ∇^H its horizontal lift to FM and $\tilde{C} : \tilde{C}(t) = (C(t), X_\alpha(t))$ a curve in FM. Then, for any tangent vector $X \in T_{C(0)}M$,*

$$\tilde{C}(X^H) = \{C(X)\}^H .$$

Proof. It suffices to note that the horizontal lift $\{C(X)\}^H$ of the vector field $C(X)$, obtained along C by parallel displacement of X with respect to ∇, satisfies the equations (6.43).□

Let $\xi = (\xi^1, \ldots, \xi^n) \in \mathbf{R}^n$; then we can consider the standard horizontal vector field $B\xi$ on FM. The following is a well known result: a curve $C(t)$ in M is a geodesic of ∇ if and only if $C(t)$ is the projection onto M of an integral curve of a standard horizontal vector field $B\xi$ on FM.

Thus, if we denote by $\tilde{C}(t)$ the integral curve of $B\xi$ and by $p = C(0)$ the initial point of $\tilde{C}(t)$, then $X = \pi_*(B\xi)_p$ is the tangent vector of the projected curve $C(t) = \pi(\tilde{C}(t))$ at the initial point $x = \pi(p)$ of $C(t)$. Since $C(t)$ is a geodesic of ∇, then $C(X)$ is the tangent vector field of $C(t)$ and, assuming that ∇ is torsion free and applying Proposition 6.2.21, we deduce $\tilde{C}(X^H) = B\xi$, because $X^H_p = (B\xi)_p$. Therefore, we have proved the following:

Proposition 6.2.22 *Let ∇ be a torsion free linear connection on M. Then, every integral curve of a standard horizontal vector field on FM is a geodesic of ∇^H.*

Covariant derivative

In order to obtain some interesting geometric consequences, we turn next to the covariant derivative with respect to ∇^H of the horizontal lift F^H of a tensor field F of type $(1,1)$ and the diagonal lift G^D of a tensor field G of type $(0,2)$.

Firstly, using (6.18), (6.29) and Proposition 6.2.15, by direct computation one obtains the following identities:

$$\text{(6.44)} \qquad \begin{cases} \left(\nabla^H_{X^H} F^H\right)(Y^H) &= 0 \,, \\ \left(\nabla^H_{X^H} F^H\right)(\lambda A) &= \lambda((\nabla_X F)^\circ A) \,, \\ \left(\nabla^H_{\lambda A} F^H\right)(\lambda B) &= 0 \,, \\ \left(\nabla^H_{\lambda A} F^H\right)(X^H) &= \lambda((T(-,FX) - FT(-,X))^\circ A) \,, \end{cases}$$

for all vector fields X, Y on M and all $A, B \in gl(n, \mathbf{R})$. Therefore

Proposition 6.2.23
(1): $\nabla^H_{X^H} F^H = 0$ *if and only if* $\nabla_X F = 0$*;*
(2): $\nabla^H F^H = 0$ *if and only if, for all vector fields* X *on* M*,* $\nabla F = 0$ *and* $T(-,FX) - FT(-,X) = 0$.

Corollary 6.2.24 *Let* ∇ *be a torsionfree linear connection. Then*

$$\nabla^H F^H = 0 \iff \nabla F = 0 \,.$$

Now, let G be a tensor field of type $(0,2)$ on M. The diagonal lift G^D of G to FM with respect to a linear connection ∇ on M is a tensor field of the same type (recall Definition 5.2.2) defined as follows. Let G_{ij} be the components of G in a chart (U, x^i) in M; then, with respect to a field of adapted coframes (η^i, η^i_α) given in (5.13), G^D is locally expressed in (FU, x^i, x^i_α) by:

$$\text{(6.45)} \qquad G^D = G_{ij}\,\eta^i \otimes \eta^j + \delta^{\alpha\beta} G_{ij}\,\eta^i_\alpha \otimes \eta^j_\beta \,.$$

Then, for all vector fields X, Y on M and all $A, B \in gl(n, \mathbf{R})$,

$$\text{(6.46)} \qquad \begin{cases} G^D(X^H, Y^H) &= \{G(X,Y)\}^V \,, \\ G^D(\lambda A, X^H) &= G^D(X^H, \lambda A) = 0 \,, \\ G^D(\lambda A, \lambda B) &= G^\circ(A,B) \,, \end{cases}$$

where $G^\circ(A,B)$ is the differentiable function on FM given by

$$\text{(6.47)} \qquad G^\circ(A,B) = \delta^{\alpha\beta} A^\alpha_\beta B^\mu_\gamma x^i_\alpha x^j_\mu G_{ji} \,,$$

with $A = (A^\alpha_\beta)$, $B = (B^\mu_\gamma)$. Obviously, $G^\circ(A,B)$ is globally defined on FM.

A straightforward computation, using the identities previously obtained and (6.45)—(6.47), leads to the following set of formulas:

$$(6.48)\quad \begin{cases} (\nabla^H_{X^H} G^D)(Y^H, Z^H) &= \{(\nabla_X G)(Y,Z)\}^V , \\ (\nabla^H_{X^H} G^D)(\lambda A, \lambda B) &= (\nabla_X G)^\circ(A,B) , \\ (\nabla^H_{X^H} G^D)(Y^H, \lambda A) &= (\nabla^H_{X^H} G^D)(\lambda A, Y^H) \\ &= (\nabla^H_{\lambda A} G^D)(Y^H, Z^H) \\ &= (\nabla^H_{\lambda A} G^D)(\lambda B, \lambda C) \;=\; 0 , \\ (\nabla^H_{\lambda A} G^D)(X^H, \lambda B) &= (G(T(X,-),-))^\circ(A,B) , \\ (\nabla^H_{\lambda A} G^D)(\lambda B, X^H) &= (G(-,T(X,-))^\circ(A,B) , \end{cases}$$

for arbitrary vector fields X, Y, Z on M, any $A, B \in gl(n, \mathbf{R})$, and where $G(T(X,-),-)$ is the tensor field of type $(0,2)$ defined by

$$G(T(X,-),-)(Y,Z) = G(T(X,Y),Z) \ .$$

From (6.48), it follows

Proposition 6.2.25
(1): $\nabla_{X^H} G^D = 0$ *if and only if* $\nabla_X G = 0$;
(2): $\nabla_{\lambda A} G^D = 0$ *if and only if* ∇ *is torsion free.*

Note that if G is a Riemannian metric on M, then G^D is a Riemannian metric on FM (for more details see Chapter 7). In particular, using Propositions 6.2.16 and 6.2.25, we have:

Theorem 6.2.26 *Let G be a Riemannian metric on M and ∇ its Levi–Civita connection. Let G^D be the diagonal lift of G to FM with respect to ∇ and ∇^H the horizontal lift of ∇. Then, ∇^H is a metric connection on FM, that is $\nabla^H G^D = 0$. Moreover, ∇^H is the Levi–Civita connection of the metric G^D if and only if the metric G is flat (or locally Euclidean).*

An interesting application of this result is the following:

Let J be a tensor field of type $(1,1)$ on M defining an almost complex structure, i.e. $J^2 = -I$, and let G be an Hermitian metric on M, that is G satisfies $G(JX, JY) = G(X,Y)$ for any vector fields X, Y on M. If ∇ is the Levi–Civita connection of G, then G^D is an Hermitian metric on FM with respect to J^H because, by virtue of (6.18) and (6.46),

$$\begin{aligned}
G^D(J^H X^H, J^H Y^H) &= G^D((JX)^H, (JY)^H) = \{G(JX, JY)\}^V \\
&= \{G(X,Y)\}^V = G^D(X^H, Y^H) , \\
G^D(J^H X^H, J^H(\lambda A))) &= G^D((JX)^H, \lambda(J^\circ A)) \\
&= G^D(X^H, \lambda A) = 0 , \\
G^D(J^H(\lambda A), J^H(\lambda B)) &= G^D(\lambda(J^\circ A), \lambda(J^\circ B)) \\
&= \delta^{\gamma\mu} A^\alpha_\gamma B^\beta_\mu x^i_\alpha x^j_\beta J^k_i J^l_j G_{kl} \\
&= \delta^{\gamma\mu} A^\alpha_\gamma B^\beta_\mu x^i_\alpha x^j_\beta G_{ij} \\
&= G^\circ(A, B) \;=\; G^D(\lambda A, \lambda B) ,
\end{aligned}$$

for all vector fields X, Y on M, all $A = (A^\alpha_\gamma)$, $B = (B^\beta_\mu) \in gl(n, \mathbf{R})$, and $\{J^j_i\}$ being the local components of J.

Hence, recalling that the almost Hermitian structure (J, G) on M is Kähler if and only if $\nabla J = 0$, combining Corollary 6.2.24 and Theorem 6.2.26, one easily deduces:

Corollary 6.2.27 *The frame bundle of a flat Kähler manifold is also a Kähler manifold.*

Canonical flat connection on FM

Let ∇ be a linear connection on M. Then, FM is a parallelizable manifold and the absolute parallelism associated to ∇ is determined by the global field of frames $\{E_\alpha, E^\mu_\gamma\}$ given in (6.2), (cf. also (8.1) below). Thus, there exists on FM a flat connection $\overline{\Gamma}$ whose covariant derivative $\overline{\nabla}$ is given as follows: let X, Y be vector fields on FM and $Y = a^\alpha E_\alpha + b^\gamma_\mu E^\mu_\gamma$; then

$$\overline{\nabla}_X Y = X(a^\alpha)\, E_\alpha + X(b^\gamma_\mu)\, E^\mu_\gamma .$$

In particular,

$$\text{(6.49)} \qquad \begin{cases} \overline{\nabla}_{E_\alpha} E_\beta = 0 , & \overline{\nabla}_{E_\alpha} E^\mu_\gamma = 0 , \\ \overline{\nabla}_{E^\gamma_\mu} E_\alpha = 0 , & \overline{\nabla}_{E^\gamma_\mu} E^\rho_\sigma = 0 . \end{cases}$$

Let U be a coordinate neighbourhood on M, Γ^h_{ji} the components of ∇ in U; then, in FU, $\{E_\alpha, E^\mu_\gamma\}$ is expressed, in terms of the local adapted frame $\{D_j, D^k_\mu\}$ given in (5.12), as follows:

$$\text{(6.50)} \qquad E_\alpha = x^j_\alpha\, D_j , \quad E^\mu_\gamma = \sum_{k=1}^n x^k_\gamma\, D^k_\mu .$$

Therefore, using (6.49) and (6.50), it follows

$$(6.51) \qquad \begin{cases} \overline{\nabla}_{D_i} D_j = \Gamma^h_{ij} D_h & , \ \overline{\nabla}_{D_i} D^j_\gamma = \sum\limits_{k=1}^{n} \Gamma^k_{ij} D^k_\gamma , \\ \overline{\nabla}_{D^i_\alpha} D_j = -x^\alpha_j D_i & , \ \overline{\nabla}_{D^i_\alpha} D^j_\beta = -x^\alpha_j D^i_\beta . \end{cases}$$

Let X be a vector field on M with local components X^i in U, $A = (A^\beta_\alpha) \in gl(n, \mathbf{R})$; then, in FU we can write

$$X^H = X^j D_j , \quad \lambda A = \sum_{k=1}^{n} \sum_{\alpha=1}^{n} A^\beta_\alpha x^k_\beta D^k_\alpha ,$$

and, through a direct computation, we obtain the following global identities, which describe the covariant derivative $\overline{\nabla}$ of the canonical flat connection on FM.

Proposition 6.2.28 *For any $A, B \in gl(n, \mathbf{R})$ and any vector fields X, Y on M,*

$$(6.52) \qquad \begin{cases} \overline{\nabla}_{X^H} Y^H = (\nabla_X Y)^H & , \ \overline{\nabla}_{\lambda A} \lambda B = 0 , \\ \overline{\nabla}_{X^H} \lambda A = 0 & , \ \overline{\nabla}_{\lambda A} X^H = -(A \cdot X)^H , \end{cases}$$

where $(A \cdot X)^H$ is the vector field on FM locally given by

$$(A \cdot X)^H = A^\alpha_\beta x^\beta_i X^i E_\alpha ,$$

$\{X^i\}$ being the local components of X and $A = (A^\beta_\alpha)$.

Remark 6.2.29 Note that, for $A = I$, $(I \cdot X)^H = X^H$.

Proposition 6.2.30 *Always $\nabla \neq \nabla^C$ and $\nabla \neq \nabla^H$.*

Moreover, taking into account (6.51) and (6.52), one easily checks the vanishing of the curvature tensor of $\overline{\nabla}$. For the torsion tensor $\overline{T}$ of $\overline{\nabla}$ one obtains:

$$\begin{aligned} \overline{T}(X^H, Y^H) &= \{T(X,Y)\}^H + \gamma R(X,Y) , \\ \overline{T}(X^H, \lambda A) &= -(A \cdot X)^H , \\ \overline{T}(\lambda A, \lambda B) &= -\lambda [A, B] . \end{aligned}$$

Corollary 6.2.31 *The canonical flat connection $\overline{\Gamma}$ is never torsion free.*

Derivations

In this section we shall keep the same notation as in Section 4.6.

Let ∇ be a linear connection on M. The following proposition is similar to Proposition 4.6.2, and it can be easily proved by similar arguments.

Proposition 6.2.32 *Two derivations D and D' of $T(FM)$ coincide if and only if:*

(a) $Df^V = D'f^V$, $Df^{(\alpha)} = D'f^{(\alpha)}$, $1 \leq \alpha \leq n$, *for any function f on M;*

(b) $DY^H = D'Y^H$ *and* $D(\lambda A) = D'(\lambda A)$, *for any vector field Y on M and any $A \in gl(n, \mathbf{R})$.*

Let $D = \mathcal{L}_X + \iota_F$ be a derivation of $T(M)$. By definition, *the horizontal lift D^H of D to FM* is given by

$$D^H = \mathcal{L}_{X^H} + \iota_{F^H} .$$

Proposition 6.2.33

(a) *For any function f on M,*

$$\begin{aligned} D^H f^V &= \{Df\}^V ; \\ D^H f^{(\alpha)} &= \{Df\}^{(\alpha)} - (\widehat{\nabla}_\gamma X) f^{(\alpha)} , \quad 1 \leq \alpha \leq n ; \\ D^H f^C &= \{Df\}^C - (\widehat{\nabla}_\gamma X) f^C . \end{aligned}$$

(b) *For any vector field Y on M and any $A \in gl(n, \mathbf{R})$,*

$$\begin{aligned} D^H Y^H &= (DY)^H - \gamma R(X, Y) , \\ D^H(\lambda A) &= \lambda(F^0 A) . \end{aligned}$$

Proof. Direct computation.□

Let $D = \mathcal{L}_X + \iota_F$, $\overline{D} = \mathcal{L}_{\overline{X}} + \iota_{\overline{F}}$ be two derivations of $T(M)$. From (6.16), (6.18), (6.19), (6.20) and Propositions 6.2.2 and 6.2.33, we have

$$\begin{aligned} &([D^H, \overline{D}^H] - [D, \overline{D}]^H) f^V = 0 , \\ &([D^H, \overline{D}^H] - [D, \overline{D}]^H) f^{(\alpha)} = -(\gamma R(X, \overline{X})) f^{(\alpha)} , \quad 1 \leq \alpha \leq n , \\ &([D^H, \overline{D}^H] - [D, \overline{D}]^H) Y^H = -\gamma\{R(X, [\overline{X}, Y]) - R(\overline{X}, [X, Y]) \\ &\qquad\qquad R(Y, [X, \overline{X}]) + R(X, \overline{F}Y) + FR(\overline{X}, Y) \\ &\qquad\qquad -R(\overline{X}, FY) - \overline{F}R(X, Y)\} , \\ &([D^H, \overline{D}^H] - [D, \overline{D}]^H)(\lambda A) = \lambda((\nabla_X \overline{F} - \nabla_{\overline{X}} F)^0 A) , \end{aligned}$$

for all functions f and vector fields Y on M, and any $A \in gl(n, \mathbf{R})$.

Then, using (6.53) and Proposition 6.2.32, we have:

Proposition 6.2.34

(**a**) *The map* $\iota_F \longmapsto \iota_{F^H}$ *is a Lie algebra homomorphism.*

(**b**) *If* ∇ *is flat, then the map* $\mathcal{L}_X \longmapsto \mathcal{L}_{X^H}$ *is a Lie algebra homomorphism.*

Remark 6.2.35 However, the map $D \longmapsto D^H$ is not a Lie algebra homomorphism even if ∇ is flat.

Next, we shall study the horizontal lifts of covariant differentiations. Let ∇_X be the covariant differentiation with respect to a vector field X on M. Then, $\nabla_X = \mathcal{L}_X + \iota_F$ where $F = \widehat{\nabla} X$.

Consider the horizontal lift $(\nabla_X)^H$; then:

$$
(6.53) \qquad \begin{cases} (\nabla_X)^H f^V &= (\nabla_X f)^V = (Xf)^V , \\ (\nabla_X)^H f^{(\alpha)} &= (\nabla_X f)^{(\alpha)} - (\widehat{\nabla}_\gamma X) f^{(\alpha)} \\ &= (Xf)^{(\alpha)} - (\widehat{\nabla}_\gamma X) f^{(\alpha)} , \ 1 \leq \alpha \leq n , \\ (\nabla_X)^H Y^H &= (\nabla_X Y)^H - \gamma R(X,Y) , \\ (\nabla_X)^H (\lambda A) &= \lambda(F^0 A) , \end{cases}
$$

for all function f and vector fields X on M, and all $A \in gl(n, \mathbf{R})$.

On the other hand, we can consider the horizontal lift ∇^H of ∇ to FM, and the covariant differentiation $\nabla^H_{X^H}$. From (6.16), (6.54) and Proposition 6.2.2, we obtain:

$$
(6.54) \qquad \begin{cases} \nabla^H_{X^H} f^V &= X^H f^V = (Xf)^V , \\ \nabla^H_{X^H} f^{(\alpha)} &= X^H f^{(\alpha)} = (Xf)^{(\alpha)} - (\widehat{\nabla}_\gamma X) f^{(\alpha)} , \\ \nabla^H_{X^H} Y^H &= (\nabla_X Y)^H , \\ \nabla^H_{X^H} \lambda A &= 0 , \end{cases}
$$

for all function f and vector fields X on M, and all $A \in gl(n, \mathbf{R})$.

Thus, bearing in mind Proposition 6.2.32, and comparing (6.54) and (6.55), we deduce:

Proposition 6.2.36 *If* ∇ *is flat, then* $(\nabla_X)^H = \nabla^H_{X^H}$ *for all parallel vector fields* X *on* M.

Chapter 7

Lift G^D of a Riemannian G to FM

Introduction

Let Γ be a linear connection on M, ∇ its covariant derivative and G a tensor field of type $(0,2)$ on M. In Chapter 5 we have introduced the diagonal lift of tensor fields on M to FM with respect to the connection Γ. Since that lift preserves the type of the tensor, we shall be interested throughout this Chapter initially in those tensors of type $(0,2)$ in general, and later in the important case of the lift of metric tensors, because these will produce Riemannian metrics on FM, and that makes their study specially interesting.

We elicit detailed properties of the curvature tensor of the diagonal lift metric on FM and study the orthonormal bundle and geodesics.

Two applications are described. We construct the lift of an almost contact structure on M to an almost Hermitian structure on FM. We find conditions for the lifted structure to be Hermitian and show that it cannot be almost Kähler nor nearly Kähler unless it is actually Kähler.

Also, we consider the lifting and projection of harmonic maps. The bundle projection is a harmonic fibration and local diffeomorphisms between manifolds have total geodesicity preserved under their prolongation; harmonicity is preserved between flat base manifolds. Thus, the diagonal lift metric on FM allows an extension of a number of results already known for the corresponding structure on TM for a Riemannian manifold (M,G).

7.1 G^D, G of type (0,2)

Let us recall that if G is a tensor field of type $(0,2)$ and ∇ is a linear connection on M, then the diagonal lift G^D of G to FM with respect to ∇ is the tensor field on FM of type $(0,2)$ with local expression in (FU, x^i, x^i_α):

$$G^D = G_{ij}\eta^i \otimes \eta^j + \delta^{\alpha\beta}G_{ij}\eta^i_\alpha \otimes \eta^j_\beta \,, \tag{7.1}$$

G_{ij} being the local components of G in (U, x^i), and (η^i, η^i_α) the local field of coframes given in (5.3). Equivalently, G^D is intrinsically defined by (6.46).

From (7.1) it easily follows that if G has constant rank r, then G^D has constant rank $(n+1)r$. Therefore:

Proposition 7.1.1

(1): *If G is a Riemannian metric on M, then G^D is a Riemannian metric on FM.*

(2): *If G is an almost symplectic differential form on M, then G^D is an almost symplectic differential form on FM.*

Remark 7.1.2 The Riemannian metric G^D on FM was first considered by K.P. Mok [64]. Section 7.2 in this Chapter will be devoted to a study of the Levi–Civita connection of G^D. G^D has been called the *Sasaki–Mok metric on FM* in [19], because of its resemblance to the Sasaki metric of the tangent bundle TM of a Riemannian manifold [79].

Next, coming back to the general situation, we shall compute the Lie derivatives of G^D with respect to vector fields λA, X^C or X^H on FM. For this, recall that the Lie derivative $\mathcal{L}_{\widetilde{X}}\widetilde{G}$ of a tensor field $\widetilde{G}$ of type $(0,2)$ with respect to a vector field $\widetilde{X}$ is given by:

$$(\mathcal{L}_{\widetilde{X}}\widetilde{G})(\widetilde{Y}, \widetilde{Z}) = \widetilde{X}(\widetilde{G}(\widetilde{Y}, \widetilde{Z}) - \widetilde{G}([\widetilde{X}, \widetilde{Y}], \widetilde{Z}) - \widetilde{G}(\widetilde{Y}, [\widetilde{X}, \widetilde{Z}]) ,$$

for all vector fields $\widetilde{Y}$, $\widetilde{Z}$; also, recall (6.47).

Proposition 7.1.3 *For all $A, B \in gl(n, \mathbf{R})$ and vector fields X, Y on M,*

(1): $(\mathcal{L}_{\lambda A}G^D)(X^H, \lambda B) = (\mathcal{L}_{\lambda A}G^D)(\lambda B, X^H) = 0$;

(2): $(\mathcal{L}_{\lambda A}G^D)(X^H, Y^H) = 0$;

(3): $(\mathcal{L}_{\lambda A}G^D)(\lambda B, \lambda C) = G^\circ(B(A + {}^tA), C)$, *where ${}^tA$ denotes the transpose of A.*

Proof. **(1)** and **(2)** follow directly from (6.46) and (6.19). For **(3)**,

$$\begin{aligned}(\mathcal{L}_{\lambda A}G^D)(\lambda B, \lambda C) = \lambda A(G^\circ(B, C)) - G^\circ(AB, C) \\ +G^\circ(BA, C) - G^\circ(B, AC) + G^\circ(B, CA)\end{aligned}$$

Using (6.47), one obtains $\lambda A(G^\circ(B, C)) = G^\circ(AB, C) + G^\circ(B, AC)$, and hence

$$(\mathcal{L}_{\lambda A}G^D)(\lambda B, \lambda C) = G^\circ(BA, C) + G^\circ(B, CA) = G^\circ(B(A + {}^tA), C) .\square$$

Corollary 7.1.4 *Let G be a Riemannian metric (resp. an almost symplectic differential form) on M. Then the fundamental vector field λA on FM is a Killing vector field (resp. an infinitesimal automorphism) of (FM, G^D) if and only if A is skew-symmetric, that is $A + {}^tA = 0$.*

A direct computation from the definitions leads to the following identities:

$$(7.2) \qquad \begin{cases} G^D(\lambda A, X^C) &= G^D(\lambda A, \widehat{\nabla}_\gamma X) , \\ G^D(X^C, \lambda A) &= G^D(\widehat{\nabla}_\gamma X, \lambda A) , \\ G^D(X^C, Y^C) &= \{G(X,Y)\}^V + G^D(\widehat{\nabla}_\gamma X, \widehat{\nabla}_\gamma Y) , \\ G^D(X^C, Y^H) &= G^D(X^H, Y^C) = \{G(X,Y)\}^V . \end{cases}$$

Proposition 7.1.5 *For all $A, B \in gl(n, \mathbf{R})$ and X, Y, Z vector fields on M,*
(1): $(\mathcal{L}_{X^C} G^D)(Y^H, Z^H) = \{(\mathcal{L}_X G)(Y, Z)\}^V$;
(2): $(\mathcal{L}_{X^C} G^D)(Y^H, \lambda A) = G^D(\gamma(R(X,Y) + \nabla_Y(\widehat{\nabla} X)), \lambda A)$,
$(\mathcal{L}_{X^C} G^D)(\lambda A, Y^H) = G^D(\lambda A, \gamma(R(X,Y) + \nabla_Y(\widehat{\nabla} X)))$;
(3): $(\mathcal{L}_{X^C} G^D)(\lambda A, \lambda B) = (\mathcal{L}_X G)^\circ(A, B)$.

Proof. Direct computation.□

Recall that the vector field X on M is an affine infinitesimal transformation of the linear connection ∇ if the Lie derivative $\mathcal{L}_X \nabla$ vanishes. The following result is well known [47,p.235].

Proposition 7.1.6 *Let ∇ be a torsion free linear connection on M. Then, a vector field X on M is an affine infinitesimal transformation of ∇ if and only if for all vector fields Y on M $R(X,Y) = -\nabla_Y(\nabla X)$.*

Thus, from Propositions 7.1.5 and 7.1.6, we deduce:

Proposition 7.1.7 *Let ∇ be a torsion free linear connection on M, G^D the diagonal lift with respect to ∇ of a tensor field G on M of type $(0,2)$ to FM, and X a vector field on M. Then, the condition*

$$\mathcal{L}_{X^C} G^D = 0$$

is equivalent to the conditions

$$\mathcal{L}_X G = 0 \quad , \quad \mathcal{L}_X \nabla = 0 .$$

Proposition 7.1.8 *For all $A, B \in gl(n, \mathbf{R})$ and X, Y, Z vector fields on M,*
(1): $(\mathcal{L}_{X^H} G^D)(Y^H, Z^H) = \{(\mathcal{L}_X G)(Y, Z)\}^V$;
(2): $(\mathcal{L}_{X^H} G^D)(Y^H, \lambda A) = G^D(\gamma R(X,Y), \lambda A)$,
$(\mathcal{L}_{X^H} G^D)(\lambda A, Y^H) = G^D(\lambda A, \gamma R(X,Y))$;
(3): $(\mathcal{L}_{X^H} G^D)(\lambda A, \lambda B) = (\nabla_X G)^\circ(A, B)$.

Proof. Direct computation.□

Corollary 7.1.9 *Let X be a vector field and G a tensor field of type* $(0,2)$ *on M. Then the condition*

$$\mathcal{L}_{X^H} G^D = 0$$

is equivalent to the conditions

$$\mathcal{L}_X G = 0 \ , \ R(X,-) = 0 \ , \ \nabla_X G = 0 \ ,$$

where $R(X,-)$ is the tensor field of type $(1,2)$ *given by*

$$R(X,-)(Y,Z) = R(X,Y)Z \ .$$

Combining all the previous results we rediscover a result by K.P. Mok [64] as follows: assume ∇ to be the Levi–Civita connection of a Riemannian metric G on M. Then:

Theorem 7.1.10 [64] *The complete (resp. horizontal) lift to FM of a vector field X on M is a Killing vector field in (FM, G^D) if and only if X is a Killing vector field (resp. a vector field with vanishing second covariant derivative) in (M,G).*

Proof. It follows immediately from Proposition 7.1.7, Corollary 7.1.9 and the following facts: if X is a Killing vector field in (M,G), i.e $\mathcal{L}_X G = 0$, then $\mathcal{L}_X \nabla = 0$ and, therefore, $R(X,Y) = -\nabla_Y(\nabla X)$. Since $(\nabla_Y(\nabla X))(Z) = (\nabla^2 X)(Z,Y)$, then $R(X,Y)Z = -(\nabla^2 X)(Z,Y)$.□

7.2 Levi–Civita connection of G^D

We begin by stating some general facts.

Let ∇ be a linear connection on M, with local components Γ^h_{ij}, and denote by

$$K: TTM \longrightarrow TM$$

the *connection map associated to* ∇ introduced by P. Dombrowski [28]. For each $\alpha = 1, 2, ..., n$, define a connection map

$$K^{(\alpha)}: TFM \longrightarrow TM$$

as follows: let $\pi^{(\alpha)}: FM \longrightarrow TM$ be the map given by $\pi^{(\alpha)}(p) = p_\alpha$, $p = (p_1, \ldots, p_n) \in FM$, and put

$$K^{(\alpha)} = K \circ T\pi^{(\alpha)} \ .$$

A direct computation leads to the following equations defining $K^{(\alpha)}$ locally:

$$\begin{aligned} K_p^{(\alpha)}\left(\frac{\partial}{\partial x^i}\right) &= \Gamma_{ij}^h(x) x_\alpha^j(p) \left(\frac{\partial}{\partial x^h}\right)_x , \\ K_p^{(\alpha)}\left(\frac{\partial}{\partial x_\beta^i}\right) &= \delta_\beta^\alpha \left(\frac{\partial}{\partial x^i}\right)_x , \end{aligned}$$

where $x = \pi(p)$.

Observe in particular that for a vector field X on M, the horizontal lift X^H and the vertical lifts $X^{(\alpha)}$ of X to FM with respect to ∇ are well defined by the following identities:

$$\begin{aligned} \pi_* X_p^H &= X_{\pi(p)} \quad , \quad K^{(\beta)} X_p^H = 0_{\pi(p)}, \\ \pi_* X_p^{(\alpha)} &= 0_{\pi(p)} \quad , \quad K^{(\beta)} X_p^{(\alpha)} = \delta^{\beta\alpha} X_{\pi(p)} , \end{aligned}$$

for every point $p \in FM$.

Let F be a tensor field on M of type $(1,1)$. For each $\alpha = 1, 2, ..., n$, we define a vertical (resp. horizontal) vector field $\gamma^\alpha F$ (resp. σF) on FM as follows. For every $p = (p_1, \ldots, p_n) \in FM$,

$$(\gamma^\alpha F)_p = \{F(p_\alpha)\}_p^{(\alpha)} \qquad (\text{ not summing over } \alpha) ; \tag{7.3}$$

$$(\sigma_\alpha F)_p = \{F(p_\alpha)\}_p^H . \tag{7.4}$$

Locally, if $\{F_j^h\}$ are the local components of F with respect to (U, x^i), then in FU:

$$\begin{aligned} \gamma^\alpha F &= x_\alpha^j F_j^h \frac{\partial}{\partial x_\alpha^j} \qquad (\text{not summing over } \alpha) , \\ \sigma_\alpha F &= F_j^h x_\alpha^j \left(\frac{\partial}{\partial x^h} - \Gamma_{ji}^h x_\beta^h \frac{\partial}{\partial x_\beta^k}\right) . \end{aligned}$$

We also denote

$$\gamma F = \sum_{\alpha=1}^n \gamma^\alpha F \quad , \quad \sigma F = \sum_{\alpha=1}^n \sigma_\alpha F . \tag{7.5}$$

In fact, γF is just the vertical vector field introduced, in a different but equivalent way, in (6.3), (6.4) and (6.5).

On the other hand, the bracket products of vertical and horizontal vector fields are given by the formulas:

$$\left\{ \begin{aligned} [X^{(\alpha)}, Y^{(\beta)}] &= 0 , \\ [X^H, Y^{(\alpha)}] &= (\nabla_X Y)^{(\alpha)} , \\ [X^H, Y^H] &= [X, Y]^H - \gamma R(X, Y) , \\ [X^{(\alpha)}, \gamma^\beta F] &= \delta^{\alpha\beta} \{F(X)\}^{(\alpha)} , \end{aligned} \right. \tag{7.6}$$

for all vector fields X, Y and all tensor field F of type $(1,1)$ on M.

Now, let (M,G) be a Riemannian manifold, ∇ an arbitrary (non metric in general) connection on M; the diagonal lift G^D of G to FM with respect to ∇ is a well defined Riemannian metric on FM (see **(1)** in Proposition 7.1.1) by the identities:

$$\begin{cases} G^D(X^H,Y^H) &= \{G(X,Y)\}^V , \\ G^D(X^H,Y^{(\alpha)}) &= 0 , \\ G^D(X^{(\alpha)},Y^{(\beta)}) &= \delta^{\alpha\beta}\{G(X,Y)\}^V , \end{cases} \tag{7.7}$$

for all vector fields X, Y on M and all $\alpha, \beta = 1,2,...,n$.

The Riemannian metric G^D is called *the induced Sasaki–Mok metric on FM.*

Next, assume that the connection ∇ on M is just the Levi–Civita connection of the metric G, R being the curvature tensor of ∇, and let $\widetilde{\nabla}$ denote the Levi–Civita connection of the induced Sasaki–Mok metric G^D on FM. Then, using the explicit formula

$$\begin{aligned} 2G^D(\widetilde{\nabla}_{\widetilde{X}}\widetilde{Y},\widetilde{Z}) &= \widetilde{X}G^D(\widetilde{Y},\widetilde{Z}) + \widetilde{Y}G^D(\widetilde{Z},\widetilde{X}) \\ &\quad -\widetilde{Z}G^D(\widetilde{X},\widetilde{Y}) + G^D([\widetilde{X},\widetilde{Y}],\widetilde{Z}) \\ &\quad +G^D([\widetilde{Z},\widetilde{X}],\widetilde{Y}) + G^D(\widetilde{X},[\widetilde{Z},\widetilde{Y}]) \end{aligned} \tag{7.8}$$

as well as (7.3)—(7.7), the following proposition is easily proved.

Proposition 7.2.1 *For arbitrary vector fields X, Y on M, and for any point $p = (p_1,\ldots,p_n) \in FM$, with $x = \pi(p)$, it follows:*

$$\widetilde{\nabla}_{X^{(\alpha)}}Y^{(\beta)} = 0 , \tag{7.9}$$

$$\widetilde{\nabla}_{X^H}Y^H = (\nabla_X Y)^H - (1/2)\gamma R(X,Y) , \tag{7.10}$$

$$(\widetilde{\nabla}_{X^{(\alpha)}}Y^H)_p = (1/2)\{R_x(p_\alpha,X_x,Y_x)\}^H_p , \tag{7.11}$$

$$(\widetilde{\nabla}_{X^H}Y^{(\alpha)})_p = (\nabla_X Y)^{(\alpha)}_p + (1/2)\{R_x(p_\alpha,Y_x,X_x)\}^H_p . \tag{7.12}$$

Since the horizontal and the vertical lifts to FM of vector fields on M generate the Lie algebra of vector fields on FM, formulas (7.9)—(7.12) above completely determine the Levi–Civita connection $\widetilde{\nabla}$ of the metric of Sasaki–Mok G^D on FM.

Remark 7.2.2 There is a similarity between the formulas we have obtained here and those obtained by O. Kowalski [49] for the Levi–Civita connection of the Sasaki metric on the tangent bundle TM of M. Moreover K.P. Mok [64] has determined the connection $\widetilde{\nabla}$ by computing its local components. In [64], Mok studies the parallelism properties for $\widetilde{\nabla}$, and his results can be easily reobtained from the formulas above. In Sections 7.7 and 7.8 we discuss applications to the lifting of almost contact structures, yielding almost Hermitian structures on FM, and to the lifting of harmonic local diffeomorphisms between Riemannian manifolds.

7.3 Curvature of G^D

In this paragraph we shall determine the curvature tensor $\widetilde{R}$ of the Levi–Civita connection $\widetilde{\nabla}$ of G^D. For that, we need an auxiliary lemma.

Lemma 7.3.1 *Let (M,G) be a Riemannian manifold, and let $\widetilde{\nabla}$ be the Levi–Civita connection of (FM,G^D). Let F be a tensor field on M of type $(1,1)$; then, for any vector field X on M and any point $p=(p_1,\ldots,p_n)\in FM$ with $x=\pi(p)$, the following assertions are true:*

(i): *for every $\alpha,\beta\in\{1,2,...,n\}$,*

$$\widetilde{\nabla}_{X^{(\alpha)}}(\gamma^\beta F)=\delta^\beta_\alpha\{F(X)\}^{(\beta)}\ , \tag{7.13}$$

(and hence $\widetilde{\nabla}_{X^{(\alpha)}}(\gamma F)=\{F(X)\}^{(\alpha)}$;)

$$\begin{aligned}(\widetilde{\nabla}_{X^{(\alpha)}}(\sigma_\beta F))_p &= \delta_{\alpha\beta}\{F(X)\}^H_p\\ &\quad+(1/2)\{R_x(p_\alpha,X_x,F(p_\beta))\}^H_p\ ;\end{aligned} \tag{7.14}$$

(ii): *if V is an arbitrary vector field on M such that $V_x=p_\alpha$ and $(\nabla_X V)_x=0$, then*

$$(\widetilde{\nabla}_{X^H}(\gamma^\alpha F))_p=(\widetilde{\nabla}_{X^H}\{F(V)\}^{(\alpha)})_p\ , \tag{7.15}$$

$$(\widetilde{\nabla}_{X^H}(\sigma_\alpha F))_p=(\widetilde{\nabla}_{X^H}\{F(V)\}^H)_p\ . \tag{7.16}$$

Proof.

(i): A direct computation, using (7.8) and the subsequent formulas in Section 7.2, leads to the identities:

$$\begin{aligned}G^D(\widetilde{\nabla}_{X^{(\alpha)}}(\gamma^\beta F),Z^H) &= 0\ ,\\ G^D(\widetilde{\nabla}_{X^{(\alpha)}}(\gamma^\beta F),Z^{(\gamma)}) &= \delta^\beta_\alpha G^D(\{F(X)\}^{(\beta)},Z^{(\gamma)})\ ,\end{aligned}$$

from where (7.13) follows directly.

The proof of (7.14) is a little more complicated. First, since ∇ is the Levi–Civita connection of G and using the formulas in Section 7.2, we have

$$G^D(\widetilde{\nabla}_{X^{(\alpha)}}(\sigma_\beta F), Z^{(\gamma)}) = 0 \ . \tag{7.17}$$

Now, we compute the horizontal component $h[X^{(\alpha)}, \sigma_\beta F]$ of $[X^{(\alpha)}, \sigma_\beta F]$, and the vertical component $v[Z^H, \sigma_\beta F]$ of $[Z^H, \sigma_\beta F]$, which are given by

$$h[X^{(\alpha)}, \sigma_\beta F] = \delta^\alpha_\beta \{F(X)\}^H \ , \tag{7.18}$$

$$v[Z^H, \sigma_\beta F] = -F^h_j x^j_\beta \gamma R\left(Z, \frac{\partial}{\partial x^h}\right) \ . \tag{7.19}$$

From (7.18) it follows

$$G^D([X^{(\alpha)}, \sigma_\beta F], Z^H) = \delta^\alpha_\beta \{G(F(X), Z)\}^V \ .$$

From (7.19), at any point $p = (p_1, \dots, p_n) \in FM$ with $x = \pi(p)$, one has:

$$\begin{aligned} G^D_p(X^{(\alpha)}, [Z^H, \sigma_\beta F]) &= G^D_p(X^{(\alpha)}, \gamma R(F(p_\beta), Z)) \\ &= G_x(X_x, R_x(F(p_\beta), Z, p_\alpha)) \\ &= G_x(R_x(p_\alpha, X, F(p_\beta)), Z_x) \\ &= G^D_p(\{R_x(p_\alpha, X, F(p_\beta)\}^H_p, Z^H_p) \ . \end{aligned}$$

Finally,

$$X^{(\alpha)} G^D(\sigma_\beta F, Z^H) = \delta^\alpha_\beta \{G(F(X), Z)\}^V \ ,$$

and, through a direct computation using again (7.8), one obtains

$$\begin{aligned} 2G^D_p(\{\widetilde{\nabla}_{X^{(\alpha)}}(\sigma_\beta F)\}_p, Z^H_p) &= 2\delta^\alpha_\beta G^D_p(\{F(X)\}^H_p, Z^H_p) \\ &\quad + G^D_p(\{R_x(p_\alpha, X, F(p_\beta))\}^H_p, Z^H_p) \ , \end{aligned}$$

which, together with (7.17), proves (7.14).

(ii): Let (U, x^i) be a local chart on M, $\{F^h_j\}$ and $\{V^i\}$ the local components of F and V, respectively; then, on (FU, x^i, x^i_α),

$$\begin{aligned} \{\widetilde{\nabla}_{X^H}(\gamma^\alpha F)\}_p &= X^H_p(F^h_i x^i_\alpha)\left(\frac{\partial}{\partial x^h}\right)^{(\alpha)}_p + (F^h_i x^i_\alpha)(p)\left(\widetilde{\nabla}_{X^H_p}\left(\frac{\partial}{\partial x^h}\right)^{(\alpha)}\right) \ , \\ \{\widetilde{\nabla}_{X^H}\{F(V)\}^{(\alpha)}\}_p &= X^H_p(F^h_i V^i)\left(\frac{\partial}{\partial x^h}\right)^{(\alpha)}_p + (F^h_i V^i)(p)\left(\widetilde{\nabla}_{X^H_p}\left(\frac{\partial}{\partial x^h}\right)^{(\alpha)}\right) \ . \end{aligned}$$

Now, if $p_\alpha(x) = p^i_\alpha(\partial/\partial x^h)_x$, that is $x^i_\alpha(p) = p^i_\alpha$, then $V^i(x) = p^i_\alpha$ because $V_x = p_\alpha$; hence $(F^h_i x^i_\alpha)(p) = (F^h_i V^i)(p)$, and

$$\begin{aligned} X^H_p(F^h_i x^i_\alpha) &= -(X^k\Gamma^i_{kj}p^j_\alpha F^h_i)(x) + p^i_\alpha X^H_p(F^h_i) , \\ X^H_p(F^h_i V^i) &= p^i_\alpha X^H_p(F^h_i) + F^h_i(x)X^H_p(V^i) , \end{aligned}$$

where $\{X^h\}$ are the local components of X in U. Therefore, in order to prove (7.15) we are reduced to proving the identity

$$X^H_p(V^i) = -(X^H\Gamma^i_{kj})(x)p^j_\alpha \quad \text{for each } \alpha . \tag{7.20}$$

To prove this identity, we first recall the hypothesis $\nabla_{X_x}V = 0$; hence, for each i,

$$\left(X^k\frac{\partial V^i}{\partial x^k}\right)(x) = -(X^h\Gamma^i_{kj})(x)p^j_\alpha ,$$

and, on the other hand, from the local expression of X^H, we have

$$X^H_p(V^i) = \left(X^k\frac{\partial V^i}{\partial x^k}\right)(x) .$$

The proof of (7.16) is completely similar, so we shall omit it for brevity.□

We can now state the main theorem of this Section.

Theorem 7.3.2 *The curvature tensor $\widetilde{R}$ of the Levi–Civita connection $\widetilde{\nabla}$ of G^D is completely determined by the following formulas:*

$$\widetilde{R}(X^{(\alpha)}, Y^{(\beta)}, Z^{(\gamma)}) = 0 ; \tag{7.21}$$

$$\left\{ \begin{aligned} \widetilde{R}(X^H, Y^{(\alpha)}, Z^{(\beta)})_p = &-(1/2)\{\delta^{\alpha\beta}R(Y,Z,X) \\ &+(1/2)R(p_\alpha, Y, R(p_\beta, Z, X)\}^H_p ; \end{aligned} \right. \tag{7.22}$$

$$\left\{ \begin{aligned} \widetilde{R}(X^{(\alpha)}, Y^{(\beta)}, Z^H)_p = &\{\delta^{\alpha\beta}R(X,Y,Z) \\ &+(1/4)R(p_\alpha, X, R(p_\beta, Y, Z)) \\ &-(1/4)R(p_\beta, Y, R(p_\alpha, X, Z))\}^H_p ; \end{aligned} \right. \tag{7.23}$$

$$\left\{ \begin{aligned} \widetilde{R}(X^H, Y^{(\alpha)}, Z^H)_p = &(1/2)\{R(X,Z,Y)\}^{(\alpha)}_p \\ &+(1/4)\sum_{\beta=1}^{n}\{R(R(p_\alpha, Y, Z), X, p_\beta)\}^{(\beta)}_p \\ &+(1/2)\{(\nabla_X R)(p_\alpha, Y, Z)\}^H_p ; \end{aligned} \right. \tag{7.24}$$

$$(7.25)\quad \begin{cases} \widetilde{R}(X^H, Y^H, Z^{(\alpha)})_p = \{R(X,Y,Z)\}_p^{(\alpha)} \\ \qquad +(1/4)\sum_{\beta=1}^{n}\{R(R(p_\alpha, Z, Y), X, p_\beta) \\ \qquad -R(R(p_\alpha, Z, X), Y, p_\beta)\}_p^{(\beta)} \\ \qquad +(1/2)\{(\nabla_X R)(p_\alpha, Z, Y) \\ \qquad -(\nabla_Y R)(p_\alpha, Z, X)\}_p^H ; \end{cases}$$

$$(7.26)\quad \begin{cases} \widetilde{R}(X^H, Y^H, Z^H)_p = (1/2)\sum_{\alpha=1}^{n}\{(\nabla_Z R)(X, Y, p_\alpha)\}_p^{(\alpha)} \\ \qquad +\{R(X,Y,Z) \\ \qquad +\sum_{\alpha=1}^{n}((1/2)R(p_\alpha, R(X,Y,p_\alpha), Z) \\ \qquad +(1/4)R(p_\alpha, R(Z,Y,p_\alpha), X) \\ \qquad +(1/4)R(p_\alpha, R(X,Z,p_\alpha), Y))\}_p^H ; \end{cases}$$

where, for conciseness, we have written X, Y, Z, *in the place of* X_x, Y_x, Z_x, *respectively, with* $x = \pi(p)$.

Proof. Let us first recall the well known formula

$$(7.27)\qquad \widetilde{R}(\widetilde{X}, \widetilde{Y}, \widetilde{Z}) = [\widetilde{\nabla}_{\widetilde{X}}, \widetilde{\nabla}_{\widetilde{Y}}]\widetilde{Z} - \widetilde{\nabla}_{[\widetilde{X},\widetilde{Y}]}\widetilde{Z} .$$

(7.21): It follows directly from (7.27), (7.6) and (7.9).
(7.22): According to (7.6) and (7.9),

$$\widetilde{R}(X^H, Y^{(\alpha)}, Z^{(\beta)}) = -\widetilde{\nabla}_{Y^{(\alpha)}}\left(\widetilde{\nabla}_{X^H} Z^{(\beta)}\right) .$$

Next, we define an auxiliary tensor field F on M of type $(1,1)$ by setting

$$F(V) = (1/2)R(V, Z_x, X_x) \text{ , for every } V \in T_x M .$$

Then, from (7.4), it follows

$$(\sigma_\beta F)_p = \{F(p_\beta)\}_p^H = (1/2)\{R(p_\beta, Z_x, X_x)\}_p^H ,$$

and then (7.12) can be written in the form

$$\widetilde{\nabla}_{X^H} Z^{(\beta)} = (\nabla_X Z)^{(\beta)} + \sigma_\beta F .$$

Since $\widetilde{\nabla}_{Y^{(\alpha)}}(\nabla_X Z)^{(\beta)} = 0$, and by (7.14), it follows

$$\begin{aligned}\widetilde{R}(X^H, Y^{(\alpha)}, Z^{(\beta)})_p &= -\{\widetilde{\nabla}_Y^{(\alpha)}(\sigma_\beta F)\}_p \\ &= -\{\delta^{\alpha\beta} F(Y_x) + (1/2)R(p_\alpha, Y_x, F(p_\beta))\}_p^H ,\end{aligned}$$

from where (7.22) follows immediately.

(7.23): It is a direct consequence of (7.22) and of the identity

$$\widetilde{R}(X^{(\alpha)}, Y^{(\beta)}, Z^H) = -\widetilde{R}(Z^H, X^{(\alpha)}, Y^{(\beta)}) + \widetilde{R}(Z^H, Y^{(\beta)}, X^{(\alpha)}) .$$

(7.24): Firstly, introduce on M an auxiliary tensor field F of type $(1,1)$ by setting

$$F(V) = (1/2)R(V, Y_x, Z_x) \ , \ \text{for any} \ \ V \in T_xM \ .$$

Then,

$$(\sigma_\alpha F)_p = (1/2)\{R(p_\alpha, Y_x, Z_x)\}_p^H ,$$

and therefore (7.11) can be written in the form

$$\widetilde{\nabla}_{Y^{(\alpha)}} Z^H = \sigma_\alpha F \ .$$

Now, using (7.11), we obtain

$$\begin{aligned}\widetilde{R}(X^H, Y^{(\alpha)}, Z^H) &= \widetilde{\nabla}_{X^H}(\sigma_\alpha F) \\ &\quad -\widetilde{\nabla}_{Y^{(\alpha)}}\left((\nabla_X Y)^H - (1/2)\gamma R(X, Z)\right) \\ &\quad -\widetilde{\nabla}_{(\nabla_X Y)^{(\alpha)}} Z^H ,\end{aligned}$$

from where (7.24) follows easily by using (7.13), (7.11) and (7.16).

(7.25): It is a direct consequence of (7.24) and of the identity

$$\widetilde{R}(X^H, Y^H, Z^{(\alpha)}) = \widetilde{R}(X^H, Z^{(\alpha)}, Y^H) - \widetilde{R}(Y^H, Z^{(\alpha)}, X^H) .$$

(7.26): According to (7.6), (7.9)—(7.11) and (7.27), one has

$$\begin{aligned}\widetilde{R}(X^H, Y^H, Z^H) &= \widetilde{\nabla}_{X^H}(\nabla_Y Z)^H - \widetilde{\nabla}_{Y^H}(\nabla_X Z)^H \\ &\quad -\widetilde{\nabla}_{[X,Y]^H} Z^H - \widetilde{\nabla}_{X^H}((1/2)\gamma R(Y, Z)) \\ &\quad +\widetilde{\nabla}_{Y^H}((1/2)\gamma R(X, Z)) + \widetilde{\nabla}_{\gamma R(X,Z)} Z^H .\end{aligned}$$

Now, we introduce once more an auxiliary tensor field F on M of type $(1,1)$ by setting

$$F(V) = -(1/2)R(Y_x, Z_x, V) \ , \ \text{for any} \ V \in T_xM \ .$$

Then, from (7.3), (7.5) and (7.15), it follows

$$\{\widetilde{\nabla}_{X^H}(-(1/2)\gamma R(Y,Z))\}_p = \sum_{\alpha=1}^{n}\{\widetilde{\nabla}_{X^H}\{F(V_\alpha)\}^{(\alpha)}\}_p ,$$

where, for each α, V_α is an arbitrary vector field on M such that $(V_\alpha)_x = p_\alpha$ and $\nabla_{X_x}V_\alpha = 0$. Then, from (7.12), it follows

$$\begin{aligned}\{\widetilde{\nabla}_{X^H}(-(1/2)\gamma R(Y,Z))\}_p &= -\sum_{\alpha=1}^{n}\Big((1/2)\{\nabla_X(R(Y,Z,V_\alpha))\}_p^{(\alpha)} \\ &\qquad +(1/4)\{R(p_\alpha,Y,Z,V_\alpha),X)\}_p^H\Big) .\end{aligned}$$

A similar argument allows us to compute

$$\{\widetilde{\nabla}_{Y^H}((1/2)\gamma R(X,Z))\}_p$$

and, using (7.9)—(7.12), the rest of the proof is routine.□

Theorem 7.3.3 [64] *(FM, G^D) is flat if and only if (M,G) is flat.*

Proof. From (7.21)—(7.26) it is clear that, if (M,G) is flat, then (FM,G^D) is also flat. Conversely, assume $\tilde{R}=0$; then, for any α,

$$4K_p^{(\alpha)}\{\tilde{R}(X^H,Y^{(\alpha)},Z^H)\}_p = R(R(p_\alpha,Y,Z),X,p_\alpha)_x + 2R(X,Z,Y)_x = 0 ,$$

for any $X,Y,Z \in T_xM$ and any $p \in FM$, with $x = \pi(p)$. In particular, taking $p \in FM$ such that $p_\alpha = Y$, then $R(X,Z,Y)_x = 0$, and hence (M,G) is flat.□

Proposition 7.3.4 *The curvature tensors R of (M,G) and $\tilde{R}$ of (FM,G^D) are related as follows:*

$$\widetilde{\nabla}\tilde{R} = 0 \implies R = 0 \implies \tilde{R} = 0 .$$

Proof. We shall compute $K_p^{(\alpha)}\{(\widetilde{\nabla}_{V^H}\tilde{R})_p(X^H,Y^{(\alpha)},Z^{(\alpha)})\}$.

For that, we begin by introducing an auxiliary tensor field F of type $(1,1)$ on M by setting

$$F(W) = R(W,Y_x,R(W,Z_x,X_x)) \ , \text{ for any } W \in T_xM .$$

Then, with the obvious simplifications in notation, we obtain

$$\begin{aligned}&(\widetilde{\nabla}_{V^H}\ \tilde{R})_p(X^H,Y^{(\alpha)},Z^{(\alpha)}) \\ &= \widetilde{\nabla}_{V_p^H}\tilde{R}(X^H,Y^{(\alpha)},Z^{(\alpha)}) - \tilde{R}(\widetilde{\nabla}_{V^H}X^H,Y^{(\alpha)},Z^{(\alpha)})_p \\ &\quad -\tilde{R}(X^H,\widetilde{\nabla}_{V^H}Y^{(\alpha)},Z^{(\alpha)})_p - \tilde{R}(X^H,Y^{(\alpha)},\widetilde{\nabla}_{V^H}Z^{(\alpha)})_p \\ &= \widetilde{\nabla}_{V_p^H}\Big(-(1/2)\{R(Y,Z,X)\}^H - (1/4)\sigma_\alpha F\Big) \\ &\quad -\tilde{R}(\{\nabla_V X\}^H - (1/2)\gamma R(V,X),Y^{(\alpha)},Z^{(\alpha)})_p \\ &\quad -\tilde{R}(X^H,\{\nabla_V Y\}^{(\alpha)} + (1/2)\{R(p_\alpha,Y,V)\}^H,Z^{(\alpha)})_p \\ &\quad -\tilde{R}(X^H,Y^{(\alpha)},\{\nabla_V Z\}^{(\alpha)} + (1/2)\{R(p_\alpha,Z,V)\}^H)_p ,\end{aligned}$$

from where, having in mind Theorem 7.3.2, we deduce by direct computations:

$$\begin{aligned}
&4\,K_p^{(\alpha)}\{(\widetilde{\nabla}_{V^H}\widetilde{R})_p(X^H, Y^{(\alpha)}, Z^{(\alpha)})\} \\
&= R(V, R(Y,Z,X), p_\alpha) \\
&\quad -R(X, R(p_\alpha, Z, V), Y) - 2R(X, R(p_\alpha, Y, V), Z) \\
&\quad +(1/2)\Big(R(V, R(p_\alpha, Y, R(p_\alpha, Z, X)), p_\alpha) \\
&\quad +R(R(p_\alpha, Z, X), R(p_\alpha, Y, V), p_\alpha) \\
&\quad -R(R(p_\alpha, Z, R(p_\alpha, Y, V)), X, p_\alpha) \\
&\quad -R(R(p_\alpha, Y, R(p_\alpha, Z, V)), X, p_\alpha))\Big)\ .
\end{aligned}$$

If $\widetilde{\nabla}R = 0$ holds, then the term in the right side above vanishes; substituting there $p_\alpha = Y$ and $p_\alpha = Z$, we obtain

$$\begin{aligned}
R(V, R(Y,Z,X), Y) - R(X, R(Y,Z,V), Y) &= 0 \\
R(V, R(Y,Z,X), Z) - 2R(X, R(Z,Y,V), Z) &= 0\ ,
\end{aligned}$$

and changing Y with Z in the second identity above:

$$R(V, R(Z,Y,X), Y) - 2R(X, R(Z,Y,V), Z) = 0\ .$$

Now, from the first and the third relations it follows:

$$R(X, R(Y,Z,V), Y) = 0\ .$$

Multiplying by a vector W and, after a simple reordering, we get

$$G(R(Y,W,X), R(Y,Z,V)) = 0\ .$$

Finally, if we put $W = Z$ and $X = V$ it follows $\|R(Y,Z,V)\| = 0$, and hence $R = 0$. Using Theorem 7.3.2 the proof is done.□

Corollary 7.3.5 [64] *(FM, G^D) is never locally symmetric unless the manifold (M,G) is locally Euclidean.*

Next, we compute the sectional, Ricci and scalar curvature of (FM, G^D). Let $p \in FM$ and $\widetilde{X}$, $\widetilde{Y}$ orthonormal vectors in p. Let $\tilde{\kappa}(\widetilde{X}, \widetilde{Y})$ be the sectional curvature of the 2–dimensional section generated by $\widetilde{X}$, $\widetilde{Y}$, that is:

$$\tilde{\kappa}(\widetilde{X}, \widetilde{Y}) = G^D(R(\widetilde{X}, \widetilde{Y}, \widetilde{Y}), \widetilde{X})\ .$$

Then,
(i): If $\widetilde{X} = X_p^H$, $\widetilde{Y} = Y_p^H$, with $X, Y \in T_xM$, $x = \pi(p)$, then

$$\begin{aligned}\tilde{\kappa}(X_p^H, Y_p^H) &= G_p^D(\tilde{R}(X^H, Y^H, Y^H)_p, X^H) \\ &= G_x(R(X,Y,Y),X) \\ &\quad +(3/4)\sum_{\alpha=1}^n G_x(R(p_\alpha, R(X,Y,p_\alpha)),Y),X) \\ &= \kappa(X,Y) - (3/4)\sum_{\alpha=1}^n \|R_x(X,Y,p_\alpha)\|^2 ,\end{aligned}$$

where $\kappa(X,Y)$ is the sectional curvature in M.
(ii): If $\widetilde{X} = X_p^H$, $\widetilde{Y} = Y_p^{(\alpha)}$, with $X, Y \in T_xM$, $x = \pi(p)$, then

$$\begin{aligned}\tilde{\kappa}(X_p^H, Y_p^{(\alpha)}) &= G_p^D(\tilde{R}(X^H, Y^{(\alpha)}, Y^{(\alpha)})_p, Y^{(\alpha)}) \\ &= -(1/4)G_x(R(p_\alpha, Y, R(p_\alpha, Y, X)), X) \\ &= (1/4)\|R(Y, p_\alpha, X)\|^2 .\end{aligned}$$

(iii): If $\widetilde{X} = X_p^{(\alpha)}$, $\widetilde{Y} = Y_p^{(\beta)}$, with $X, Y \in T_xM$, $x = \pi(p)$, then

$$\tilde{\kappa}(X_p^{(\alpha)}, Y_p^{(\beta)}) = 0 .$$

Summing up

$$\begin{aligned}\tilde{\kappa}(X_p^H, Y_p^H) &= \kappa(X,Y) - (3/4)\sum_{\alpha=1}^n \|R_x(X,Y,p_\alpha)\|^2 ; \\ \tilde{\kappa}(X_p^H, Y_p^{(\alpha)}) &= (1/4)\|R_x(Y, p_\alpha, X)\|^2 ; \\ \tilde{\kappa}(X_p^{(\alpha)}, Y_p^{(\beta)}) &= 0 .\end{aligned}$$

Thus we can state the following theorem.

Theorem 7.3.6 *If (FM, G^D) has bounded sectional curvature, then it is flat.*

Proof. Assume FM is non flat. Then, by virtue of Theorem 7.3.3, M is non flat too; therefore, there exist a point $x \in M$ and orthonormal tangent vectors $X, Y \in T_xM$ such that $R_x(X,Y,p_\alpha) \neq 0$ for some $p = (p_1, \ldots, p_n) \in FM$ with $x = \pi(p)$ and some α. Hence

$$\tilde{\kappa}(X_p^H, Y_p^H) = \kappa(X,Y) - (3/4)\sum_{\alpha=1}^n \|R_x(X,Y,p_\alpha)\|^2 .$$

Since the set of $p \in FM$ satisfying such a condition is not bounded, then

$\tilde{\kappa}(X_p^H,, Y_p^H)$ has no lower bound. Using a similar procedure it can be verified that $\tilde{\kappa}(X_p^H, Y_p^{(\alpha)})$ has no upper bound, which ends the proof.□

Let $\widetilde{S}$ be the Ricci curvature of (FM, G^D). Then, if $\{V_i\}$ is an orthonormal basis for T_xM, $x = \pi(p)$, $\{(V_i^{(\alpha)})_p, (V_i^H)_p\}$ is an orthonormal basis for T_pFM. Hence, for all $\widetilde{X}, \widetilde{Y} \in T_pFM$,

$$\widetilde{S}_p(\widetilde{X}, \widetilde{Y}) = \sum_{i=1}^{n}\sum_{\alpha=1}^{n} G_p^D(\tilde{R}_p(V_i^{(\alpha)}, \widetilde{X}, \widetilde{Y}), (V_i^{(\alpha)})_p) + \sum_{i=1}^{n} G_p^D(\tilde{R}_p(V_i^H, \widetilde{X}, \widetilde{Y}), (V_i^H)_p) ,$$

from which, by direct computations, we obtain

$$\widetilde{S}_p(X_p^{(\alpha)}, Y_p^{(\beta)}) = \frac{1}{4}\sum_{i=1}^{n} G_x(R_x(p_\alpha, X, V_i), R_x(p_\beta, Y, V_i)) \; ; \tag{7.28}$$

$$\widetilde{S}_p(X_p^H, Y_p^{(\alpha)}) = \frac{1}{4}\sum_{i=1}^{n} G_x((\nabla_{V_i} R)_x(p_\alpha, Y, X), V_i) \; ; \tag{7.29}$$

$$\begin{aligned} \widetilde{S}_p(X_p^H, Y_p^H) &= S(X,Y) \\ &+(1/4)\sum_{i=1}^{n}\sum_{\alpha=1}^{n}\Big(G_x(R_x(V_i, p_\alpha, X), R_x(V_i, p_\alpha, Y)) \\ &-3G_x(R_x(V_i, X, p_\alpha), R_x(V_i, Y, p_\alpha))\Big) , \end{aligned} \tag{7.30}$$

for all $X, Y, \in T_xM$, $x = \pi(p)$, and S being the Ricci curvature tensor of M.

Theorem 7.3.7 *If (FM, G^D) is an Einstein manifold, then (M, G) is flat.*

Proof. Assume that (FM, G^D) is an Einstein manifold; then there exists a constant λ such that $\widetilde{S} = \lambda G^D$. Therefore, for all $p \in FM$, $X, Y \in T_xM$ with $x = \pi(p)$, $\alpha \in \{1, 2, ..., n\}$, one has

$$\widetilde{S}_p(X_p^{(\alpha)}, Y_p^{(\alpha)}) = \lambda G_x(X, Y) \; . \tag{7.31}$$

In particular, if we take $p = (p_1, \dots, p_n) \in FM$ such that $p_\alpha = X$, then from (7.28) and (7.31) we obtain $0 = \lambda G_x(X, Y)$ for arbitrary X, Y, so $\lambda = 0$ and hence $\widetilde{S}$ vanishes identically. Thus, setting $\alpha = \beta$ and $X = Y$ in (7.28), we get

$$0 = \sum_{i=1}^{n} \|R_x(p_\alpha, X, V_i)\|^2 ,$$

and therefore $R_x(p_\alpha, X, V_i) = 0$ for each i. Hence $R = 0$.□

Let $\tilde{r}$ be the scalar curvature of (FM, G^D); if $\{V_i\}$ is an orthonormal basis for T_xM as above, then for any $p \in FM$ with $x = \pi(p)$, $\tilde{r}(p)$ is given by

$$\tilde{r}(p) = \sum_{i=1}^{n}\sum_{\alpha=1}^{n} \tilde{S}_p((V_i^{(\alpha)})_p, (V_i^{(\alpha)})_p) + \sum_{i=1}^{n} \tilde{S}_p((V_i^H)_p, (V_i^H)_p) .$$

Then, from (7.28)—(7.30), we obtain

$$\tilde{r}(p) = r(x) + (1/2) \sum_{i,j,k} \|R_x(V_i, p_\alpha, V_j)\|^2 - (3/4) \sum_{i,j,k} \|R_x(V_j, V_i, p_\alpha)\|^2 , \tag{7.32}$$

where r denotes the scalar curvature of (M, G). In view of this formula, we can state the following

Theorem 7.3.8 *(FM, G^D) and (M, G) have the same constant scalar curvature if and only if both are flat, and then $\tilde{r} = r = 0$.*

Proof. If $\tilde{r} = r = c$, then taking $p = \{V_i\}$ in (7.32), it follows

$$\|R_x(V_i, V_j, V_k)\| = 0$$

for every i,j,k, and hence $R = 0$. The converse is trivial.□

7.4 Bundle of orthonormal frames

Let OM be the subset of FM of all orthonormal frames at every point of M. OM is a $(1/2)n(n+1)$-dimensional submanifold of FM, and, also, a reduced subbundle of FM with structural group $O(n, \mathbf{R})$, the orthogonal group. OM is called the principal bundle of orthonormal frames of M and, in what follows, we study the properties of OM as a submanifold of the Riemannian manifold (FM, G^D). The results in this section are due to K.P. Mok [64].

Let (U, x^i) be a local chart on M, and let OU be the restriction of OM to U, that is $OU = FU \cap OM$. Thus, OU is an open submanifold of OM determined by the equations

$$G_{ij} x^i_\alpha x^j_\beta = \delta_{\alpha\beta} . \tag{7.33}$$

Now, for each pair of indices α, β, we consider the differentiable function

$$f_{\alpha\beta} = G_{ij} x^i_\alpha x^j_\beta - \delta_{\alpha\beta} ,$$

and the (local) differential 1-form in FU given by

$$df_{\alpha\beta} = \frac{\partial G_{ij}}{\partial x^k} x^i_\alpha x^j_\beta dx^k + (G_{ik} x^i_\alpha \delta^\gamma_\beta + G_{kj} x^j_\beta \delta^\gamma_\alpha) dx^k_\gamma \ . \tag{7.34}$$

From (7.34) it follows easily that $df_{\alpha\beta}$ vanishes on T_pOM, for every $p \in OU$, when we are restricted to OU.

As is well known [47], there is a vector field $\nabla f_{\alpha\beta}$ on OU, canonically associated to the 1-form $df_{\alpha\beta}$, given by

$$< Y, (df_{\alpha\beta})_p >= G^D_p((\nabla f_{\alpha\beta})_p, Y) \ , \tag{7.35}$$

for every $Y \in T_pFM$, $p \in FM$.

After a simple computation from (7.34) and (7.35), we easily deduce the local expression of $\nabla f_{\alpha\beta}$ on FU:

$$\nabla f_{\alpha\beta} = (x^k_\alpha \delta_{\beta\gamma} + x^k_\beta \delta_{\alpha\gamma}) \frac{\partial}{\partial x^k_\gamma} \ . \tag{7.36}$$

If we now consider $G^D(\nabla f_{\alpha\beta}, \nabla f_{\alpha'\beta'})$ for two pairs of indices (α, β), (α', β'), and noting that the orthogonal complement of T_pOM in T_pFM has dimension $(1/2)n(n+1)$, we obtain:

Proposition 7.4.1 [64] *The collection*

$$\{(1/2)\nabla f_{\alpha\alpha} \ ; 1 \leq \alpha \leq n\} \cup \{(1/\sqrt{2})\nabla f_{\alpha\beta} \ ; 1 \leq \alpha, \beta \leq n\}$$

is a basis of the orthogonal complement of OU.

Next, we shall use Proposition 7.4.1 to find a condition for a vector field on FM to be tangent to OM. Let $p \in OU$ and $\widetilde{X} \in T_pFM$. Since $\{\nabla f_{\alpha\beta} \ ; \alpha \leq \beta\}$ is a family of normal and linearly independent vectors tangent to OU, and $\nabla f_{\alpha\beta} = \nabla f_{\beta\alpha}$, it follows that $\widetilde{X}$ is tangent to OM if and only if $G^D(\widetilde{X}, \nabla f_{\alpha\beta}) = 0$ for every α and β. If the local components of $\widetilde{X}$ are $\{\widetilde{X}^h, \widetilde{X}^h_\gamma\}$, then the following proposition follows easily:

Proposition 7.4.2 [64] *Let $p = (x^i, x^i_\alpha) \in OU$ and $\widetilde{X} = (\widetilde{X}^h, \widetilde{X}^h_\gamma) \in T_pFM$. Then $\widetilde{X}$ is tangent to OU if and only if*

$$(\widetilde{X}^a \Gamma^h_{ad} x^d_\gamma + \widetilde{X}^d_\gamma) x^\beta_h + (\widetilde{X}^a \Gamma^h_{cd} x^d_\gamma + \widetilde{X}^h_\beta) x^\gamma_h = 0 \ , \tag{7.37}$$

where (x^α_i) denotes the inverse matrix of (x^i_α).

Now we can use Proposition 7.4.2 to determine a condition for the complete lift of a vector field to be tangent to OM. For that we need a lemma.

Lemma 7.4.3 [64] *Let X be a vector field on M. Then X is a Killing vector field in (M,G) if and only if $G(A_XY,Z)+G(Y,A_XZ)=0$ for all vector fields Y, Z on M, A_X denoting the tensor field of type $(1,1)$ on M defined by $A_X=\mathcal{L}_X-\nabla_X$.*

Proof. Since $\nabla_X G=0$ for every vector field X, then $\mathcal{L}_X G=0$ is equivalent to $\nabla_X G=0$. But A_X is a derivation on the algebra of tensor fields, so

$$A_X(G(Y,Z))=(A_XG)(Y,Z)+G(A_XY,Z)+G(Y,A_XZ)$$

for all vector fields Y, Z on M. Since A_X acts trivially on functions,

$$A_X(G(Y,Z))=0\ .$$

In consequence, $A_XG=0$ if and only if $G(A_XY,Z)+G(Y,A_XZ)=0$.□

Theorem 7.4.4 [64] *Let X be a vector field on M. Then, the complete lift X^C of X to FM is tangent to OM if and only if X is a Killing vector field in (M,G).*

Proof. In FU, X^C has local components given by (1.3). Then, condition (7.37) can be written as

$$\left(X^j\Gamma^h_{ji}x^i_\gamma+x^i_\gamma\frac{\partial x^h}{\partial x^i}\right)G_{hk}x^k_\beta+\left(X^j\Gamma^h_{ji}x^i_\beta+x^i_\beta\frac{\partial x^h}{\partial x^i}\right)G_{hk}x^k_\gamma=0\ ,$$

because at points of OM one has $G_{ij}x^i_\alpha x^j_\beta=\delta_{\alpha\beta}$, and thus $x^\beta_h=G_{hk}x^k_\beta$.

In its turn, this condition can be written as

$$\left(X^j\Gamma^h_{ji}+\frac{\partial x^h}{\partial x^i}\right)x^i_\gamma x^k_\beta G_{hk}+\left(X^j\Gamma^h_{ji}+\frac{\partial x^h}{\partial x^i}\right)x^i_\beta x^k_\gamma G_{hk}=0\ ,$$

or equivalently,

$$(\nabla_iX^h)G_{hk}x^i_\gamma x^k_\beta+(\nabla_kX^h)G_{hi}x^i_\gamma x^k_\beta=0\ ,$$

where ∇_iX^h is the component of the covariant derivative $\nabla_{\partial/\partial x^i}X$. Then, it follows that the condition (7.37) for X^C is equivalent to

$$((\nabla_iX^h)G_{hk}+(\nabla_kX^h)G_{hi})x^i_\gamma x^k_\beta=0\ ,$$

and, since this expression is invariant under changes of orthonormal frames, it is equivalent to

$$(\nabla_iX^h)G_{hk}+(\nabla_kX^h)G_{hi}=0\ .$$

But the last identity is equivalent to the identity

$$G(A_XY,Z)+G(Y,A_XZ)=0\ ,$$

and the result follows using Lemma 7.4.3.□

The proof of Theorem 7.4.4 can be found also in [47].

Remark 7.4.5 If (M,G) is an almost symplectic manifold, it is possible to develop a similar study for the bundle of almost symplectic frames SM as submanifold of FM. For the details of such a study we refer the reader to [12].

7.5 Geodesics of G^D

Our purpose in this paragraph is to state some relations between the geodesics of (FM, G^D) and those of (M, G).

The first result is the following.

Proposition 7.5.1 [64] *Let σ be a curve in M and $\tilde{\sigma}$ its horizontal lift to FM. Then $\tilde{\sigma}$ is a geodesic in (FM, G^D) if and only if σ is a geodesic in (M, G).*

Proof. Since $\dot{\tilde{\sigma}} = (\dot{\sigma})^H$, using (7.10), it follows

$$\widetilde{\nabla}_{\dot{\tilde{\sigma}}}\dot{\tilde{\sigma}} = (\nabla_{\dot{\sigma}}\dot{\sigma})^H$$

and the assertion is proved.□

Now, recall the well known result [47]:

Proposition 7.5.2 *The projection onto M of any integral curve of a standard horizontal vector field on FM is a geodesic of (M, G) and, conversely, every horizontal lift of a geodesic in (M, G) is an integral curve of a standard horizontal vector field on FM.*

Taking into account that a curve in FM is horizontal if and only if its tangent vector at any point is horizontal, from Propositions 7.5.1 and 7.5.2 it follows

Proposition 7.5.3 [64] *A curve in FM is a horizontal geodesic of G^D if and only if it is an integral curve of some standard horizontal vector field.*

Remark 7.5.4 Proposition 7.5.2 characterizes the geodesics of G on M in terms of standard horizontal vector fields on FM. There is, however, another characterization of the geodesics in M in terms of *the geodesic spray* S in M: a curve $x^i = x^i(t)$ is a geodesic in (M, G) if and only if its natural lift to TM, $x^i = x^i(t)$, $y^i = dx^i/dt$, is an integral curve of S. Thus, the standard horizontal vector fields on FM play the role of S.

7.6 Applications

f–structures on FM

The natural structure as a principal fibre bundle, which is inherent to the frame bundle of every differentiable manifold, allows the construction of some canonical f-structures on FM when M is a Riemannian manifold.

Let (M, G) be a Riemannian manifold of dimension n. Define a tensor field F_α of type $(1,1)$ on FM, for each $\alpha = 1, 2, \ldots, n$, by

$$(7.38) \qquad F_\alpha X^H = -X^{(\alpha)}\ , \ F_\alpha X^{(\beta)} = \delta_\alpha^\beta X^H\ ,$$

for all vector fields X on M, where the horizontal lifts are considered with respect to the Levi–Civita connection of G.

Each F_α has constant rank $2n$ and $F_\alpha^3 + F_\alpha = 0$, so it defines an f–structure of rank $2n$ over FM (see also Section 8.3).

The integrability and the partial integrability of an f–structure were first studied by Ishihara–Yano [43]. In order to study them for the f–structure F_α on FM, we need the following lemma.

Lemma 7.6.1 *Let H be a tensor field of type $(1,1)$ on M. Then:*
(1) $F_\alpha(\gamma H) = \sigma_\alpha H$;
(2) $F_\alpha(\sigma_\alpha H) = -\gamma^\alpha H$.

Proof. Direct from (7.3), (7.4) and (7.5).□

Let $l_\alpha = -F_\alpha^2$, $m_\alpha = F_\alpha^2 + I$ be the projection operators associated to F_α, and let $\mathcal{L}_\alpha = Im\ l_\alpha$, $\mathcal{M}_\alpha = Im\ m_\alpha$ be the complementary distributions on FM associated to l_α and m_α, respectively.

Then $dim\ \mathcal{L}_\alpha = 2n$ and $dim\ \mathcal{M}_\alpha = n^2 - n$. Moreover,

$$(7.39) \qquad \begin{cases} l_\alpha(X^H) = X^H\ , & l_\alpha(X^{(\beta)}) = \delta_\alpha^\beta X^{(\beta)}, \\ m_\alpha(X^H) = 0\ , & m_\alpha(X^{(\beta)}) = (I - \delta_\alpha^\beta) X^{(\beta)}\ , \end{cases}$$

Then $\{X^H, X^{(\alpha)}\}$ span $\mathcal{L}_\alpha$ and $\{X^{(\beta)}; \beta \neq \alpha\}$ span $\mathcal{M}_\alpha$.

From Lemma 7.6.1, it follows that $\mathcal{M}_\alpha$ is always completely integrable; the integrability of $\mathcal{L}_\alpha$ is determined in the following proposition.

Proposition 7.6.2 *$\mathcal{L}_\alpha$ is completely integrable if and only if (M, G) is locally Euclidean.*

The Nijenhuis tensor of F_α behaves as follows:

$$(7.40) \qquad \begin{cases} N_{F_\alpha}(X^H, Y^H) = \gamma^\alpha R(X, Y)\ , \\ N_{F_\alpha}(X^H, Y^{(\beta)}) = \delta_\alpha^\beta \sigma_\alpha R(X, Y)\ , \\ N_{F_\alpha}(X^{(\beta)}, Y^{(\mu)}) = -\delta_\alpha^\beta \delta_\alpha^\mu \gamma R(X, Y)\ , \end{cases}$$

for all vector fields X, Y, Z on M, and $1 \leq \beta \leq n$.

From Proposition 7.6.2 and (7.40), it follows easily:

Proposition 7.6.3 *The following assertions are equivalent:*
(1): *(M, G) is locally Euclidean;*
(2): *$\mathcal{L}_\alpha$ is completely integrable;*
(3): *F_α is partially integrable;*
(4): *F_α is integrable.*

Now, let us recall some definitions and results of the theory of f-structures. Let F be an f-structure on a manifold N, and let G be a Riemannian metric on N adapted to F. Let Ω be the fundamental 2-form of F, defined by $\Omega(X,Y) = G(FX,Y)$ for all vector fields X, Y on N. Let ∇ be the Levi-Civita connection of G. Vorha and Singh [84] have introduced the following terminology, by paralleling the one usual in Hermitian geometry; N is called:
(1): *fAK-manifold* if and only if $d\Omega(FX, FY, FZ) = 0$;
(2): *fH-manifold* if and only if F is partially integrable;
(3): *fK-manifold* if and only if $\nabla_{FX}F = 0$, for all vector fields X on N.
Then,

Proposition 7.6.4 [84] *N is an fK-manifold if and only if it is both an fAK-manifold and an fH-manifold.*

From a geometric point of view, note that N is an fK-manifold if and only if the distribution $\mathcal{L}$ is completely integrable and each of its integral manifolds is Kählerian.

Let us go back to the frame bundle (FM, G^D) of a Riemannian manifold (M, G). Then,

Proposition 7.6.5 *G^D is adapted to F_α.*

Proof. In fact the distributions $\mathcal{L}_\alpha$ and $\mathcal{M}_\alpha$ are mutually orthogonal with respect to G^D, and $G^D(\widetilde{X}, F_\alpha\widetilde{X}) = 0$, for all vector fields $\widetilde{X}$ on FM.□

The fundamental 2-form Ω_α of F_α is given by

$$\Omega_\alpha(\widetilde{X}, \widetilde{Y}) = G^D(F_\alpha\widetilde{X}, \widetilde{Y})$$

for all vector fields $\widetilde{X}$, $\widetilde{Y}$ on FM. Therefore,

$$(7.41)\qquad \begin{cases} \Omega_\alpha(X^H, Y^H) & = 0\,, \\ \Omega_\alpha(X^H, Y^{(\beta)}) & = -\delta^\beta_\alpha\{G(X,Y)\}^V\,, \\ \Omega_\alpha(X^{(\beta)}, Y^{(\mu)}) & = 0\,, \end{cases}$$

for all vector fields X, Y on M, $1 \leq \beta, \mu \leq n$.

A simple computation from (7.41) shows that $d\Omega_\alpha = 0$. Therefore, we have

Theorem 7.6.6

(1) *FM is always an fAK-manifold;*

(2) *FM is an fK-manifold if and only if (M,G) is locally Euclidean.*

Remark 7.6.7 Observe that an obvious modification in (7.38) allows us to define on FM a tensor field F'_α of type $(1,1)$, $1 \leq \alpha \leq n$, such that $F'^3_\alpha - F'_\alpha = 0$, and with *rank* $F'_\alpha = 2n$. These $f(3,-1)$-structures on FM are all different from that considered by Okubo in [74] (see also the examples in Section 8.3). We omit the study of these structures on FM because it is similar to that of F_α above.

Almost Hermitian structure

A $(2n+1)$-dimensional manifold M is said to have a (ϕ,ξ,η)-structure or an almost contact structure if it admits a tensor field ϕ of type $(1,1)$, a vector field ξ, and a 1-form η satisfying

$$\phi^2 = -I + \eta \otimes \xi \quad , \quad \eta(\xi) = 1 \, .$$

Then $\phi(\xi) = 0$ and $\eta \circ \phi = 0$. A manifold M with a (ϕ,ξ,η)-structure and a Riemannian metric G such that

$$G(\phi X, \phi Y) = G(X,Y) - \eta(X)\eta(Y) \, ,$$

for all vector fields X, Y on M, is said to have a (ϕ,ξ,η,G)-structure or an almost contact metric structure, and G is called a compatible metric. Note that $\eta(X) = G(X,\xi)$. The 2-form Φ on M defined by

$$\Phi(X,Y) = G(X,\phi Y)$$

is called the fundamental form of the almost contact metric structure. (We refer the reader to [6] for an introduction to the geometry of almost contact metric manifolds.)

Let us now suppose that M is a manifold endowed with an almost contact metric structure, and let ∇ denote the Levi-Civita connection of G.

Let J be the tensor field on FM of type $(1,1)$ defined by

$$J = \phi^H + \sum_{\alpha=1}^{n} \eta^{H_\alpha} \otimes \xi^{(\alpha+n)} - \sum_{\alpha=1}^{n} \eta^{H_{\alpha+n}} \otimes \xi^{(\alpha)} + \eta^V \otimes \xi^{(2n+1)} - \eta^{H_{2n+1}} \otimes \xi^H \, ,$$

where (see (6.15)) if $\eta = \eta_i dx^i$ is the local expression of η in (U,x^i), then

$$\begin{aligned} \eta^V &= \eta_i dx^i \, ; \\ \eta^{H_\alpha} &= x^j_\alpha \Gamma^h_{ij} \eta_h dx^i + \eta_i dx^i_\alpha \, . \end{aligned}$$

Theorem 7.6.8 [7] *(FM, G^D, J) is an almost Hermitian manifold.*

Proof. Since the horizontal and the vertical lifts of vector fields on M generate the Lie algebra of vector fields on FM, to prove that $J^2(\widetilde{X}) = -\widetilde{X}$ for all vector fields $\widetilde{X}$ on FM, it suffices to check the identity for $\widetilde{X} = X^{(\beta)}$, $\beta = 1, 2, ..., 2n+1$, and $\widetilde{X} = X^H$, X being an arbitrary vector field on M. Then:

$$J(X^{(\beta)}) = (\phi X)^{(\beta)} + \varepsilon(\beta)\eta(X)^V \xi^{(\beta+\varepsilon(\beta)n)} - \delta^\beta_{2n+1}\eta(X)^V \xi^H ,$$

where

$$\varepsilon(\beta) = \begin{cases} 1 & \text{if} \quad \beta \leq n , \\ -1 & \text{if} \quad n < \beta \leq 2n , \\ 0 & \text{if} \quad \beta = 2n+1 , \end{cases}$$

and

$$J(X^H) = (\phi X)^H + \eta(X)^V \xi^{(2n+1)} .$$

To prove that G^D is Hermitian, we consider the following three particular cases:

(1): $\widetilde{X} = X^H$, $\tilde{Y} = Y^H$. Since G is adapted to the almost contact metric structure on M, the identity $G^D(JX^H, JY^H) = G^D(X^H, Y^H)$ follows easily from (7.7).

(2): $\widetilde{X} = X^{(\beta)}$, $\tilde{Y} = Y^{(\mu)}$, $\beta, \mu = 1, 2, ..., 2n+1$. In this case

$$\begin{aligned} G^D(JX^{(\beta)}, JY^{(\mu)}) = \; & \delta^\beta_\mu \{G(\phi X, \phi Y)\}^V \\ & + \varepsilon(\beta)\varepsilon(\mu)\delta^{\beta+\varepsilon(\beta)n}_{\mu+\varepsilon(\mu)n}\eta(X)^V\eta(Y)^V \\ & + \delta^\beta_{2n+1}\delta^\mu_{2n+1}\eta(X)^V\eta(Y)^V , \end{aligned}$$

and since $\delta^{\beta+\varepsilon(\beta)n}_{\mu+\varepsilon(\mu)n} = \delta^\beta_\mu$, we get $\varepsilon(\beta)\varepsilon(\mu)\delta^\beta_\mu = \varepsilon(\beta)^2\delta^\beta_\mu$, and therefore

$$\begin{aligned} G^D(JX^{(\beta)}, JY^{(\mu)}) = & \; \delta^\beta_\mu \left(\{G(\phi X, \phi Y)\}^V + \eta(X)^V\eta(Y)^V\right) \\ = & \; G^D(X^{(\beta)}, Y^{(\mu)}) . \end{aligned}$$

(3): $\widetilde{X} = X^{(\beta)}$, $\tilde{Y} = Y^H$, $\beta = 1, 2, \ldots, 2n+1$.
In this case $G^D(X^{(\beta)}, Y^H) = 0$ and

$$\begin{aligned} G^D(JX^{(\beta)},\, JY^H) = G^D(&(\phi X)^{(\beta)} + \varepsilon(\beta)\eta(X)^V\xi^{(\beta+\varepsilon(\beta)n)} \\ & -\delta^\beta_{2n+1}\eta(X)^V\xi^H, (\phi Y)^H + \eta(Y)^V\xi^{(2n+1)}) = 0 . \square \end{aligned}$$

The following propositions are proved by direct computations from the definitions.

Proposition 7.6.9 [7] *The Kähler form F of (G^D, J) on FM is given by*

$$\begin{aligned} F(X^H, Y^H) &= \{\Phi(X,Y)\}^V , \\ F(X^H, Y^{(\beta)}) &= -\delta^{\beta}_{2n+1}\eta(X)^V\eta(Y)^V , \\ F(X^{(\beta)}, Y^{(\mu)}) &= \delta^{\beta}_{\mu}\{\Phi(X,Y)\}^V + \varepsilon(\mu)\delta^{\beta}_{\mu+\varepsilon(\mu)n}\eta(X)^V\eta(Y)^V , \end{aligned}$$

for all X, Y vector fields on M, and $\beta, \mu = 1, 2, \ldots, 2n+1$.

Proposition 7.6.10 [7] *The differential dF is given by*

$$\begin{aligned} dF(X^H, Y^H, Z^H) &= \{d\Phi(X,Y,Z)\}^V \\ &\quad + (1/3) \sum_{X,Y,Z} F(\gamma R(X,Y), Z^H) , \\ dF(X^H, Y^H, Z^{(\beta)}) &= (1/3)F(\gamma R(X,Y), Z^{(\beta)}) \\ &\quad + (1/3)\delta^{\beta}_{2n+1}\{\eta(X)(\nabla_Y\eta)(Z) \\ &\quad - \eta(Y)(\nabla_X\eta)(Z) - 2d\eta(X,Y)\eta(Z)\}^V , \\ dF(X^H, Y^{(\alpha)}, Z^{(\beta)}) &= (1/3)\{\delta^{\beta}_{\alpha}(\nabla_X\Phi)(Y,Z) \\ &\quad + \varepsilon(\alpha)\delta^{\beta}_{\alpha+\varepsilon(\alpha)n}[(\nabla_X\Phi)(Y,\phi Z) \\ &\quad - (\nabla_X\Phi)(\phi Y, Z)]\}^V , \\ dF(X^{(\alpha)}, Y^{(\beta)}, Z^{(\mu)}) &= 0 , \end{aligned}$$

for all vector fields X, Y, Z on M, and where $\sum_{X,Y,Z}$ means cyclic sum.

In order to obtain the coderivative δF of the Kähler form F, observe that if $\{E_i, \phi E_i, \xi; i = 1, 2, \ldots, n\}$ is a ϕ–basis for (M, ϕ, ξ, η, G) (see [6]), then

$$\{E_i^H, (\phi E_i)^H, \xi^H, E_i^{(\alpha)}, (\phi E_i)^{(\alpha)}, \xi^{(\alpha)}\}$$

$i = 1, 2, \ldots, n$, $\alpha = 1, 2, \ldots, 2n+1$ is a J–basis for (FM, G^D, J), that is, a basis adapted to the almost Hermitian structure on FM.

Proposition 7.6.11 [7] *The coderivative δF is given by*

$$\begin{aligned} \delta F(X^{(\beta)}) &= -\sum_{i=1}^{n}\{G^D(\gamma R(E_i, \phi E_i), X^{(\beta)}\} \\ &\quad +\delta^{\beta}_{2n+1}(\eta(X)^V(\delta\eta)^V + (\nabla_\xi\eta)(X)^V) , \\ \delta F(X^H) &= \{(\delta\Phi)(X)\}^V , \end{aligned}$$

for every vector field X on M.

In order to relate the normality of the structure (ϕ, ξ, η) (see [6]) with the integrability of the almost complex structure J on FM, we firstly determine the Nijenhuis torsion N_J of J.

Proposition 7.6.12 [7] *The Nijenhuis torsion N_J of the almost complex structure J on FM is given by*

(a):

$$\begin{aligned} N_J(X^H, Y^H) = &\{([\phi,\phi] + 2d\eta \otimes \xi)(X,Y)\}^H \\ &+\{(\mathcal{L}_{\phi X}\eta)Y - (\mathcal{L}_{\phi Y}\eta)X\}^V \xi^{(2n+1)} \\ &-\eta(X)^V\{\nabla_{\phi Y}\xi - \phi\nabla_Y\xi\}^{(2n+1)} \\ &+\eta(Y)^V\{\nabla_{\phi X}\xi - \phi\nabla_X\xi\}^{(2n+1)} \\ &-\gamma R(\phi X, \phi Y) + J(\gamma R(\phi X, Y)) \\ &+J(\gamma R(X, \phi Y)) + \gamma R(X,Y) \ ; \end{aligned}$$

(b):

$$\begin{aligned} N_J\ (X^{(\alpha)}, Y^{(\beta)}) = &\ \delta^\beta_{2n+1}\Big\{\eta(Y)^V\{\nabla_\xi(\phi)X\}^{(\alpha)} \\ &+\eta(X)^V\eta(Y)^V(\nabla_\xi\xi)^{(\alpha+\varepsilon(\alpha)n)} + \varepsilon(\alpha)\eta(Y)^V\{(\nabla_\xi\eta)\}^V\xi^{(\alpha+\varepsilon(\alpha)n)}\Big\} \\ &-\delta^\alpha_{2n+1}\Big\{\eta(X)^V\{\nabla_\xi(\phi)Y\}^{(\beta)} + \eta(X)^V\eta(Y)^V\{\nabla_\xi\xi\}^{(\beta+\varepsilon(\beta)n)} \\ &+\varepsilon(\beta)\eta(X)^V\{(\nabla_\xi\eta)Y\}^V\xi^{(\beta+\varepsilon(\beta)n)}\Big\} \\ &+\delta^\alpha_{2n+1}\delta^\beta_{2n+1}\{\eta(X)^V\{(\nabla_\xi\eta)Y\}^V - \eta(Y)^V\{(\nabla_\xi\eta)X\}^V\}\xi^H \ ; \end{aligned}$$

for every $\alpha, \beta = 1, 2, \ldots, 2n+1$.

(c): *if $\beta \neq 2n+1$,*

$$\begin{aligned} N_J(X^H, Y^{(\beta)}) = &\{(\nabla_{\phi X}\phi)Y + (\nabla_X\phi)\phi Y - \eta(Y)(\nabla_X\xi)\}^{(\beta)} \\ &+\varepsilon(\beta)\eta(Y)^V\{\nabla_{\phi X}\xi - \phi\nabla_X\xi\}^{(\beta+\varepsilon(\beta)n)} \\ &+\varepsilon(\beta)\{(\phi X)(\eta(Y)) - \eta(\nabla_{\phi X}Y) - \eta(\nabla_X\phi Y)\}^V\xi^{(\beta+\varepsilon(\beta)n} \ ; \end{aligned}$$

(d):

$$\begin{aligned} N_J(X^H, Y^{(2n+1)}) = &\{(\nabla_{\phi X}\phi)Y + (\nabla_X\phi)\phi Y - \eta(Y)(\nabla_X\xi)\}^{(2n+1)} \\ &+\{\eta(\nabla_{\phi X}Y) + \eta(\nabla_X\phi Y) - (\phi X)\eta(Y)\}^V\xi^H \\ &+\{\eta(Y)(\mathcal{L}_\xi\phi)X\}^V + \eta(Y)^V(\gamma R(\phi X, \xi) \\ &-J(\gamma R(X,\xi)) + \eta(X)^V\eta(Y)^V(\nabla_\xi\xi)^{(2n+1)} \\ &+\{\eta(Y)(\mathcal{L}_\xi\eta)(X)\}^V\xi^{(2n+1)} \end{aligned}$$

where X and Y are arbitrary vector fields on M.

Proposition 7.6.13 [7] *If the structure (ϕ, ξ, η, G) on M is normal and satisfies the $K_{1\phi}$-curvature identity, then (FM, G^D, J) is a Hermitian manifold.*

Proof. Since (M, ϕ, ξ, η, G) is normal, it follows $[\phi, \phi] + 2d\phi \otimes \xi = 0$, which implies $(\mathcal{L}_{\phi X}\eta)Y - (\mathcal{L}_{\phi Y}\eta)X = 0$, and

$$(\nabla_X \phi)Y - (\nabla_{\phi X}\phi)\phi Y + \eta(Y)\nabla_{\phi X}\xi = 0 \ . \tag{7.42}$$

This, together with the $K_{1\phi}$-curvature identity

$$R(X, Y, Z, W) = R(X, Y, \phi Z, \phi W)$$

leads to $N_J(X^H, Y^H) = 0$.

Using again (7.42), we obtain

$$(\nabla_\xi \phi)X = 0 \ , \ \nabla_\xi \xi = 0 \ , \ (\nabla_\xi \eta)(X) = G(\nabla_\xi \xi, X) = 0 \ ,$$

and thus $N_J(X^{(\alpha)}, Y^{(\beta)}) = 0$ for all vector fields X, Y on M, $\alpha, \beta = 1, 2, \ldots, 2n+1$.

Next, replacing X by ϕX in (7.42),

$$(\nabla_{\phi X}\phi)Y + (\nabla_X \phi)\phi Y - \eta(Y)(\nabla_X \xi) = 0 \ ,$$

and applying η to (7.42), a direct computation leads to

$$(\phi X)(\eta(Y)) - \eta(\nabla_{\phi X} Y) - \eta(\nabla_X \phi Y) = 0 \ .$$

Finally, bearing in mind that if M is normal we also have

$$\mathcal{L}_\xi \eta = \mathcal{L}_\xi \phi = 0 \ ,$$

we may conclude $N_J(X^H, Y^{(\beta)}) = 0$, for all $\beta = 1, 2, \ldots, 2n+1$.□

Recall that a (ϕ, ξ, η, G) structure on M is said to be *cosymplectic* [6] if it is normal with Φ and η closed. A straightforward computation proves that a cosymplectic manifold satisfies the $K_{1\phi}$-curvature identity. Examples of such manifolds are provided by $M \times \mathbf{R}$, M being any Kähler manifold.

Proposition 7.6.14 [7] *If the structure (ϕ, ξ, η, G) on M is cosymplectic then (FM, G^D, J) is Hermitian.*

Remark 7.6.15 Since M is Hermitian and satisfies the K_1-curvature identity, then $M \times \mathbf{R}$ has a canonically associated normal structure satisfying the $K_{1\phi}$-curvature identity. Hence a family of manifolds for which Theorem 7.6.13 applies is $\widetilde{M} = M^* \times \mathbf{R}^5$, where M^* is an ordinary minimal surface contained in a 3-dimensional subspace [41]. Observe that these manifolds admit no cosymplectic structure, since $M^* \times \mathbf{R}^4$ is a Hermitian non Kählerian manifold [41].

Corollary 7.6.16 *(FM, G^D, J) is Kähler if and only if (M, ϕ, ξ, η, G) is flat cosymplectic.*

Proof. If M is cosymplectic then FM is Hermitian. Then, using Proposition 7.6.10, we may conclude that $dF = 0$, i.e. FM is almost Kähler and the result follows.□

Conversely, if FM is Kähler, then $dF(X^H, Y^{(\alpha)}, Z^{(\alpha)}) = 0$ for all vector fields X, Y, Z on M and $\alpha = 1, 2, \ldots, 2n+1$. Thus $(\nabla_X \phi)(Y, Z) = 0$, i.e M is a cosymplectic manifold. Since in this kind of manifold $\nabla \eta = d\eta = 0$, it follows that $R = 0$ because of Proposition 7.6.10.

Theorem 7.6.17 [7] *If (FM, G^D, J) is Hermitian , then (M, ϕ, ξ, η, G) is normal.*

Proof. Since $N_J(X^H, \xi^{(\alpha)}) = 0$ for $\alpha \neq 2n+1$, then $(\nabla_{\phi X} \phi)\xi = \nabla_X \xi$ and $(\nabla_X \phi)\xi = -\phi(\nabla_{\phi X}(\phi)\xi) = -\nabla_{\phi X}\xi$.

Taking this into account, and the fact that $N_J(X^H, (\phi Y)^{(\alpha)}) = 0$ for $\alpha \neq 2n+1$, one gets $(\nabla_{\phi X} \phi) - (\nabla_X \phi)Y - \eta(Y)\nabla_{\phi X}\xi = 0$, which is equivalent to M being normal.□

Theorem 7.6.18 *(FM, G^D, J) can be neither almost Kähler nor nearly Kähler unless it is Kähler.*

Proof. If FM were almost Kähler, then

$$dF(X^H, Y^{(\alpha)}, Z^{(\alpha)}) = (\nabla_X \phi)(Y, Z) = 0$$

for all vector fields X, Y, Z on M and $\alpha = 1, 2, \ldots, 2n+1$. Thus, M would be a cosymplectic manifold, according to Proposition 7.6.13. If FM were a nearly Kähler manifold

$$0 = G^D(\widetilde{\nabla}_{\xi^{(2n+1)}}(J)\xi^{(2n+1)}, Y^H) = (1/2)G^D(\gamma R(Y, \xi), \xi^{(2n+1)})$$

for all vector fields Y on M. Then

$$\begin{aligned} 0 &= G^D(\widetilde{\nabla}_{X^{(2n+1)}}(J)\xi^{(2n+1)}, Y^H) + G^D(\widetilde{\nabla}_{\xi^{(2n+1)}}(J)X^{(2n+1)}, Y^H) \\ &= (1/2)G^D(\gamma R(Y, \xi), X^{(2n+1)}) \ . \end{aligned}$$

Bearing in mind these expressions, if FM were nearly Kähler we would obtain

$$\begin{aligned} 0 &= G^D(\widetilde{\nabla}_{X^H}(J)Y^{(2n+1)}, Z^{(2n+1)}) + G^D(\widetilde{\nabla}_{Y^{(2n+1)}}(J)X^H, Z^{(2n+1)}) \\ &= \{G(\nabla_X \phi)(Y, Z)\}^V \end{aligned}$$

for all vector fields X, Y, Z on M, and therefore M would be a cosymplectic manifold, which together with Proposition 7.6.13 leads to the result.□

Harmonic frame bundle maps

A map $f\colon (M,G) \longrightarrow (N,G')$ between Riemannian manifolds has second fundamental form ∇df: f is called *harmonic* if the trace of ∇df is identically zero, and *totally geodesic* if ∇df itself is identically zero. For background material on harmonic maps see Eells and Lemaire [29,30]. Dodson and Vázquez-Abal [27] have extended to (FM, G^D) a number of results proved by Sanini [78] for the tangent bundle.

Lemma 7.6.19 *The chart expressions for ∇df and its trace $\tau(f)$, with respect to coordinates (x^i) on M and (x^γ) on N are:*

$$[\nabla df]^\gamma_{ij} = \frac{\delta^2}{\delta x^i \delta x^j} f^\gamma - {}^M\Gamma^k_{ij} \frac{\delta f^\gamma}{\delta x^k} + {}^N\Gamma^\gamma_{\alpha\beta} \frac{\delta f^\alpha}{\delta x^i} \frac{\delta f^\beta}{\delta x^j} ,$$

$$[\tau(f)]^\gamma = G^{ij}(\nabla df)^\gamma_{ij} .$$

From these expressions we deduce the following

Proposition 7.6.20 *The frame bundle projection, $\pi\colon FM \longrightarrow M$, is harmonic, and totally geodesic if and only if (M,G) is flat.*

Proof. By direct computation and substitution in the Lemma

$$(\widetilde{\nabla} d\pi_M)^k = \begin{pmatrix} 0 & -\frac{1}{2}R^k_{ilj}x^l_\alpha \\ -\frac{1}{2}R^k_{jli}x^l_\alpha & 0 \end{pmatrix} . \square$$

Next we consider local diffeomorphisms $f\colon M \longrightarrow N$, they induce frame bundle morphisms $Ff\colon FM \longrightarrow FN$.

Theorem 7.6.21 *Let $f\colon (M,G) \longrightarrow (N,G')$ be a local diffeomorphism of Riemannian manifolds. Then*

$$\pi_N \circ Ff = f \circ \pi_M\colon FM \longrightarrow N$$

is harmonic if and only if f is harmonic. Moreover, if Ff is harmonic but f is not harmonic then

(i): *Neither (N,G') nor (FN, G'^D) is flat.*

(ii): *(FN, G'^D) is not an Einstein manifold.*

(iii): *(FN, G'^D) does not have bounded sectional curvature.*

Proof. Direct computation shows that π_M is actually a harmonic Riemannian submersion, then the necessary and sufficient condition follows from the work of Smith [81]. From Proposition 7.6.20 and the result of Mok [64] that (N,G') is flat if and only if (FN, G'^D) is flat (cf. Theorem 7.3.3) we get **(i)**. Applying this and Theorem 7.3.7 we get **(ii)**; similarly, Theorem 7.3.6 gives **(iii)**.□

Further details can be found in [27], together with sufficient conditions for $f \circ \pi_M$ to be constant and the proof that, for compact M, f is totally geodesic if Ff is harmonic. Quite generally, Ff is harmonic if and only if Tf is harmonic and Ff is totally geodesic if and only if f is totally geodesic.

Chapter 8

Constructing G–structures on FM

Introduction

This Chapter is devoted to describe and study a procedure that allows us to construct a great variety of G–structures on FM. The inspiring idea of this procedure comes from a paper of J.M. Terrier [83], who constructed an almost complex structure on the frame bundle FM of an even dimensional manifold M endowed with a linear connection, by "modelling" the structure on the canonical almost complex structures of $\mathbf{R}^n$ and $gl(n, \mathbf{R})$, $n = 2m$. In [17] Terrier's idea has been developed and exploited in a more general context, which will be described in this Chapter. In particular we consider polynomial structures connection–induced on FM by a tensor field of type $(1,1)$ on M, so obtaining a convenient presentation of examples from Okubo [73,74]. Next we consider G–structures induced by riemannian and symplectic structures in the presence of a connection.

8.1 ω–associated G–structures on FM

Assume given a linear connection on M, ω being its connection form.

Recall that, for each $\xi \in \mathbf{R}^n$, $B\xi$ denotes the standard horizontal vector field on FM associated to ξ, and it is uniquely determined at each point $p \in FM$ as the unique horizontal vector in p such that $\pi((B\xi)_p) = p(\xi)$, p being interpreted as a linear isomorphism

$$p\colon \mathbf{R}^n \longrightarrow T_{\pi(p)}M \ .$$

As in the preceding chapters, for each $A \in gl(n, \mathbf{R})$, λA will denote the fundamental vector field associated to A.

Let us denote by

$$e = \{e_i; i = 1, 2, \ldots, n\} \ \ , \ \ E = \{E_i^j; i, j = 1, 2, ..., n\}$$

the canonical basis of $\mathbf{R}^n$ and $gl(n, \mathbf{R})$ respectively; then, the family of vector fields

$$Be = \{Be_i; i = 1, 2, ..., n\} \ , \ \lambda E = \{\lambda E_i^j; i, j = 1, 2, ..., n\}$$

defines an absolute parallelism on FM associated to the linear connection ω. This parallelism provides a trivialization of the frame bundle FFM of the manifold FM in the following way: firstly, we define a global section

$$\sigma_\omega: FM \longrightarrow FFM$$

$$p \longmapsto \sigma_\omega(p) = \tilde{p}_0$$

where $\tilde{p}_0$ is the linear frame of T_pFM given by

$$\tilde{p}_0 = \{(Be)_p, (\lambda E)_p\} = \{(Be_i)_p, (\lambda E_k^j)_p; i, j, k = 1, 2, \ldots, n\} \ .$$

Then, the trivialization

$$\tau: FM \times Gl(n + n^2, \mathbf{R}) \longrightarrow FFM \ ,$$

associated to the parallelism, is given by

$$\tau(p, A) = \sigma_\omega(p)A = \tilde{p}_0 A \ , \ p \in FM \ , \ A \in Gl(n + n^2, \mathbf{R}) \ .$$

Thus, for each Lie subgroup $G \subset Gl(n+n^2, \mathbf{R})$ we can consider the associated G–structure on FM, defined by the principal bundle

$$P_G = \tau(FM \times G) = \{p \in FFM \,|\, \tilde{p} = \tilde{p}_0 A, \ A \in G\} \ .$$

Definition 8.1.1 *P_G will be called the ω–associated G–structure on FM.*

Let $\tilde{V}$ be a (real) vector space of finite dimension; let

$$\tilde{\rho}: Gl(n + n^2, \mathbf{R}) \longrightarrow Gl(\tilde{V})$$

be a linear representation, and let $G_{\tilde{u}} \subset Gl(n + n^2, \mathbf{R})$ be the isotropy subgroup of $\tilde{u} \in \tilde{V}$. Define

$$\begin{aligned} \tilde{t}: FFM &\longrightarrow \tilde{V} \\ \tilde{p} &\longmapsto \tilde{\rho}(A^{-1})\tilde{u} \ , \end{aligned} \tag{8.1}$$

where $A \in Gl(n + n^2, \mathbf{R})$ is the unique element such that $\tilde{p} = \tilde{p}_0 A$.

Then $\tilde{t}$ is a differentiable tensor on FFM of type $(\tilde{\rho}, \tilde{V})$ with values in $\tilde{V}_{\tilde{u}} = \{\tilde{\rho}(A)\tilde{u} \,|\, A \in Gl(n + n^2, \mathbf{R})\}$ and, hence, $\tilde{t}^{-1}(\tilde{u}) \subset FFM$ is a principal subbundle with structure group $G_{\tilde{u}}$. Moreover, $\tilde{t}^{-1}(\tilde{u}) = P_{G_{\tilde{u}}}$ and the following theorem is proved.

Theorem 8.1.2 *The ω–associated $G_{\tilde{u}}$–structure on FM is actually determined by the differentiable tensor $\tilde{t}$ given in* (8.1).

In the following sections of this chapter we shall be interested in the construction and study of some particular cases of this general situation: more precisely, we shall be concerned with those G–structures on FM defined by tensor fields of types $(1, 1)$ or $(0, 2)$.

8.2 Defined by (1,1)–tensor fields

Let us consider

$$\widetilde{V} = (\mathbf{R}^{n+n^2})^* \otimes \mathbf{R}^{n+n^2} = Hom(\mathbf{R}^{n+n^2}, \mathbf{R}^{n+n^2}) ,$$

and let $\tilde{\rho}: Gl(n+n^2, \mathbf{R}) \longrightarrow Gl(\widetilde{V})$ be the canonical linear representation.

We shall make use again of the one–to–one correspondence between tensor fields $\widetilde{J}$ of type $(1,1)$ with

$$\text{(8.2)} \qquad \begin{array}{rcl} \widetilde{J}_p: T_pFM & \longrightarrow & T_pFM \\ X & \longmapsto & \tilde{p} \circ \tilde{t}(\tilde{p}) \circ \tilde{p}^{-1}(X) , \quad \tilde{p} \in \pi^{-1}(p) , \end{array}$$

and (isotropy groups) $G_{\tilde{u}}$–structures defined (cf. (3.4)) by $\widetilde{V}_{\tilde{u}}$–valued tensor fields $\tilde{t}$ of type $(\tilde{\rho}, \widetilde{V})$ on FFM.

Among all the G–structures on FM that can be defined by tensor fields of type $(1,1)$, we shall be interested in those that can be constructed from (or "modeled" over) some special structures on the "model" vector spaces $\mathbf{R}^n$ and $gl(n,\mathbf{R})$, and that are at the same time ω–associated in the sense of Section 8.1.

For that, we consider the canonical isomorphisms of vector spaces

$$\mathbf{R}^{n+n^2} \equiv \mathbf{R}^n \times \mathbf{R}^{n^2} \simeq \mathbf{R}^n \times gl(n,\mathbf{R}) .$$

We put

$$\begin{array}{rcl} V & = & (\mathbf{R}^n)^* \otimes \mathbf{R}^n = Hom(\mathbf{R}^n, \mathbf{R}^n) , \\ V' & = & gl(n,\mathbf{R})^* \otimes gl(n,\mathbf{R}) = Hom(gl(n,\mathbf{R}), gl(n,\mathbf{R})) , \end{array}$$

and denote the canonical linear representations by

$$\rho: Gl(n,\mathbf{R}) \longrightarrow Gl(V) \quad , \quad \rho': Gl(n^2,\mathbf{R}) \longrightarrow Gl(V') .$$

If

$$j: Gl(n,\mathbf{R}) \times Gl(n^2,\mathbf{R}) \longrightarrow Gl(n+n^2,\mathbf{R})$$

denotes the canonical injection, then $\tilde{\rho} \circ j = \rho \oplus \rho'$. Therefore, if $u \in V$ and $u' \in V'$, and if we take $\tilde{u} = u + u' \in \widetilde{V}$, then

$$V_u \oplus V'_{u'} \subset \widetilde{V}_{\tilde{u}} \quad \text{and} \quad j(G_u \times G_{u'}) \subset G_{\tilde{u}} .$$

From Theorem 8.1.2, it follows that, given a linear connection ω on M, the ω–associated $G_{\tilde{u}}$–structure $P_{G_{\tilde{u}}}$ on FM is determined by a differentiable tensor $\tilde{t}: FFM \longrightarrow \widetilde{V}$ of type $(\tilde{\rho}, \widetilde{V})$ and $\widetilde{V}_{\tilde{u}}$–valued. The next theorem describes its associated tensor field $\widetilde{J}$ on FM.

Theorem 8.2.1 *Under the hypothesis above,*

$$\tilde{J}_p(X) = (Bu(\theta_p(X)))_p + (\lambda u'(\omega_p(X)))_p \ ,$$

for every $X \in T_pFM$ *and all* $p \in FM$*; that is,*

$$\tilde{J} = B \circ u \circ \theta + \lambda \circ u' \circ \omega \ . \tag{8.3}$$

Proof. From (8.2), and because $\tilde{p}_0 = \sigma_\omega(p) \in \pi^{-1}(p)$,

$$\tilde{J}_p(X) = \tilde{p}_0 \circ \tilde{t}(\tilde{p}_0) \circ \tilde{p}_0^{-1}(X) \ .$$

But $\tilde{p}_0 = \{(Be)_p, (\lambda E)_p\}$ as an isomorphism of vector spaces

$$\tilde{p}_0 \colon \mathbf{R}^{n+n^2} \simeq \mathbf{R}^n \times gl(n, \mathbf{R}) \longrightarrow T_pFM$$

is given by

$$\tilde{p}_0(\xi, 0) = (B\xi)_p \ , \ \tilde{p}_0(0, A) = (\lambda A)_p \ , \ \xi \in \mathbf{R}^n \ , \ A \in gl(n, \mathbf{R}) \ .$$

From this the result follows easily bearing in mind the following identities:

$$\tilde{t}(\tilde{p}_0) = \tilde{u} = u + u' \ ,$$

$$\theta_p((Be_i)_p) = e_i \ , \quad \omega_p((Be_i)_p) = 0 \ , \ i = 1, 2, \ldots, n \ ,$$

$$\theta_p((\lambda E^i_j)_p) = 0 \ , \quad \omega_p((\lambda E^i_j)_p = E^i_j \ , \ i, j = 1, 2, \ldots, n \ . \square$$

Obviously, the structure defined by $\tilde{J}$ on FM depends on the linear connection ω; if $\omega' = \omega + \eta$ is another linear connection on M, η being a tensorial form of type $(ad, gl(n, \mathbf{R}))$, the associated tensor field $\tilde{J}'$ is given by

$$\tilde{J}' = \tilde{J} + \lambda \circ (u' \circ \eta - \eta \circ B \circ u \circ \theta) \ .$$

This is easily checked by direct computation using

$$B'\xi = B\xi - \lambda(\eta(B\xi)) \ , \ \xi \in \mathbf{R}^n \ ,$$

$$\omega \circ \lambda + B \circ \theta = 1 \ .$$

Let us also remark that the tensor field $\tilde{J}$ is 0-deformable in the sense of Lehmann-Lejeune [50], since for any point $p \in FM$ there is a linear frame $\tilde{p}_0$ at p such that $\tilde{J}_p$ is expressed with respect to $\tilde{p}_0$ by the square matrix

$$\begin{pmatrix} u & 0 \\ 0 & u' \end{pmatrix}$$

which does not depend on p. Moreover, we deduce that

$$rank \ \tilde{J}_p = rank \ u + rank \ u' \ .$$

Then, from the results in [50], it follows

Theorem 8.2.2 *The ω-associated $G_{\tilde{u}}$-structure on FM defined by $\tilde{J}$ is integrable if and only if its first structure tensor vanishes identically.*

Recall, once more, that the Nijenhuis torsion N_ϕ of a tensor field ϕ of type $(1,1)$ is given by

$$N_\phi(X,Y) = [\phi X, \phi Y] - \phi[\phi X, Y] - \phi[X, \phi Y] + \phi^2[X,Y] .$$

It is also well known that if ϕ defines on the manifold a 0-deformable G-structure, then the vanishing of N_ϕ is a necessary, but in general not sufficient, condition for the integrability of the structure [50, Th.III, p.385]. This fact justifies the determination of $N_{\tilde{J}}$; for that, the following definition will be useful.

Definition 8.2.3 *For each $\alpha \in \Lambda^2(FM, V)$, a 2-form on FM with values in a vector space V, and $J_V: V \longrightarrow V$ being an endomorphism of V, we define $\tilde{\alpha} \in \Lambda^2(FM, V)$ by*

$$\tilde{\alpha}(X,Y) = 2(-\alpha(\tilde{J}X, \tilde{J}Y) + J_V\alpha(\tilde{J}X, Y) + J_V\alpha(X, \tilde{J}Y) - J_V^2\alpha(X,Y)) ,$$

X, Y being arbitrary vector fields on FM.

Remark that, in general, $\tilde{\alpha}$ could be zero and α not zero. (See Terrier [83] and Remark 8.3.5 bellow.)

Theorem 8.2.4 *Let ω be a linear connection on M and let $\tilde{J}$ be the tensor field given by (8.3) defining the ω-associated $G_{\tilde{u}}$-structure that we are considering on FM. Then*

$$N_{\tilde{J}} = B \circ \widetilde{d\theta} + \lambda \circ \widetilde{d\omega} , \tag{8.4}$$

where $\widetilde{d\theta} \in \Lambda^2(FM, \mathbf{R}^n)$ and $\widetilde{d\omega} \in \Lambda^2(FM, gl(n,\mathbf{R}))$ are defined with respect to $u \in V$ and $u' \in V'$ respectively.

Proof. Exploiting the absolute parallelism induced by ω on FM, it suffices to compute $N_{\tilde{J}}$ in the following three cases:

(a): $X = B_1$, $Y = B_2$, $B_i = B\xi_i$, ξ_1, ξ_2 being linearly independent in $\mathbf{R}^n$;
(b): $X = B\xi$, $\xi \neq 0$, $Y = \lambda A$, $A \neq 0$, $\xi \in \mathbf{R}^n$, $A \in gl(n,\mathbf{R})$;
(c): $X = \lambda A_1$, $Y = \lambda A_2$, A_1, A_2 being linearly independent in $gl(n,\mathbf{R})$.

Now, recall the structure equations of ω:

$$d\theta = -\omega \wedge \theta + \Theta , \quad d\omega = -\omega \wedge \omega + \Omega ,$$

where Θ (resp. Ω) denotes the torsion form (resp. the curvature form) of ω. Using these, we obtain

$$[B_1, B_2] = -2\lambda\Omega(B_1, B_2) - 2B\Theta(B_1, B_2) .$$

On the other hand,

$$[B\xi, \lambda A] = -BA(\xi) \ , \ [\lambda A_1, \lambda A_2] = \lambda[A_1, A_2] \ ,$$

and the result follows easily by direct computations bearing in mind that $\tilde{J}(B\xi) = Bu(\xi)$ (because $\theta \circ B = 1$), and that $\tilde{J}(\lambda A) = \lambda u'(A)$ (because $\omega \circ \lambda = 1$).□

In fact, (8.4) can be equivalently written as

$$N_{\tilde{J}} = B \circ (\tilde{\Theta} - \widetilde{\omega \wedge \theta}) + \lambda \circ (\tilde{\Omega} - \widetilde{\omega \wedge \omega}) \ . \tag{8.5}$$

Now, assume that M possesses a G_u-structure defined by a tensor field J_M of type $(1,1)$ given by

$$(J_M)_x(X) = p \circ t(p) \circ p^{-1}(X) \ , \ X \in T_xM \ , p \in \pi^{-1}(x) \ .$$

Here $t: FM \longrightarrow V$ is a differentiable tensor of type (ρ, V) and V_u-valued, $P = t^{-1}(u) \subset FM$ the G_u-fibre bundle determining the G_u-structure on M. If X is a vector field on M, and X^H is its horizontal lift to FM with respect to ω, then:

Theorem 8.2.5 *For all vector fields X on M,*

$$\tilde{J}X^H = (J_M X)^H \ .$$

Proof. For an arbitrary point $p \in FM$, by definition of $\tilde{J}$, we have:

$$(\tilde{J}X^H)_p = (Bu\theta_p(X^H))_p = Y \ ,$$

where Y is the unique horizontal vector at p such that

$$\pi(Y) = p \circ u \circ \theta_p(X^H) = p \circ u \circ p^{-1}(X_{\pi(p)}) \ .$$

On the other hand, $(J_M X)^H = Z$, Z being the unique horizontal vector at p such that $\pi(Z) = J_M(X_{\pi(p)})$. Hence

$$\begin{aligned} Y = Z \quad &\Longleftrightarrow \quad p \circ u \circ p^{-1}(X_{\pi(p)}) = J_M(X_{\pi(p)}) \\ &\Longleftrightarrow \quad p \circ u \circ p^{-1} = J_M \\ &\Longleftrightarrow \quad p \in P \ .\Box \end{aligned}$$

Next, assume that ω is adapted to the G_u-structure P, or equivalently, that J_M is parallel with respect to ω. Then:

Theorem 8.2.6 *Under the hypothesis above if ω is adapted to the G_u-structure P on M, then on P*

$$B\Theta(X^H, Y^H) = \{N_{J_M}(X, Y)\}^H \ .$$

Proof. In accord with (8.5),

$$N_{\tilde{J}}(X^H,Y^H) = B(\tilde{\Theta}(X^H,Y^H) - \widetilde{\omega\wedge}\theta(X^H,Y^H)) + \lambda\left(\tilde{\Omega}(X^H,Y^H) - \widetilde{\omega\wedge}\omega(X^H,Y^H)\right) .$$

We are interested only in the first term on the right hand of that identity; if, moreover, we denote by h the horizontal projection with respect to ω, using the identity $h \circ \tilde{J} = \tilde{J} \circ h$ and the definition of $\widetilde{\omega\wedge}\theta$, it follows easily that $\widetilde{\omega\wedge}\theta(X^H,Y^H) = 0$. Hence

$$hN_{\tilde{J}}(X^H,Y^H) = B\tilde{\Theta}(X^H,Y^H) .$$

On the other hand

$$N_{\tilde{J}}(X^H,Y^H) = [\tilde{J}X^H,\tilde{J}Y^H] - \tilde{J}[\tilde{J}X^H,Y^H] - \tilde{J}[X^H,\tilde{J}Y^H] + \tilde{J}^2[X^H,Y^H] ,$$

and, if we restrict to P, Theorem 8.2.5 implies, together with the fact that ω is adapted, that the horizontal part of this equation above is

$$h\left(N_{\tilde{J}}(X^H,Y^H)\right)_p = [J_MX,J_MY]^H - (J_M[J_MX,Y])^H - (J_M[X,J_MY])^H + (J_M^2[X,Y])^H ,$$

which is exactly $\{N_{J_M}(X,Y)\}^H$.□

This theorem has two interesting and simple applications, both being well known results of the theory of G–structures defined by tensor fields of type $(1,1)$.

Corollary 8.2.7 *If there exists on* M *a torsion free linear connection* ω *that leaves* J_M *parallel, then* N_{J_M} *vanishes identically.*

Proof. $\Theta = 0$ implies $\tilde{\Theta} = 0$ and, by Theorem 8.2.6, $N_{J_M} = 0$.□

Corollary 8.2.8 *If* ω *is a linear connection on* M *that leaves* J_M *parallel, and if* T *denotes the torsion tensor of* ω*, then*

$$(8.6)\qquad \left\{ \begin{array}{l} N_{J_M}(X,Y) = -T(J_MX,J_MY) + J_MT(J_MX,Y) \\ \qquad\qquad + J_MT(X,J_MY) - J_M^2T(X,Y) , \end{array} \right.$$

for all vector fields X, Y *on* M.

Proof. This identity is just the projection onto M of $N_{\tilde{J}}$ evaluated on the horizontal lifts to P. In fact, it is a direct consequence of Theorem 8.2.6, the definition of T as $T_x(X,Y) = p(2\Theta(X^H,Y^H))$, $p \in \pi^{-1}(x)$, and the definition of $\tilde{\Theta}$ from Definition 8.2.3 using $u \in V$.□

Observe that (8.6) justifies Definition 8.2.3.

8.3 Application to polynomial structures on FM

Among all the G–structures that can be defined on M by a tensor field ϕ of type $(1,1)$, the so called *polynomial structures* have always gained special attention in the literature. In this section, in accordance with the general construction described in the two preceding sections, we shall study the existence and some properties of those polynomial structures that can be defined on FM from a linear connection ω on M and from algebraic "models" on $\mathbf{R}^n$ and $gl(n,\mathbf{R})$.

Definition 8.3.1 *A polynomial structure of degree k on M is a tensor field ϕ of type $(1,1)$ and constant rank, satisfying the equation*

$$Q(\phi)=\phi^k+\mathrm{a}_k\phi^{k-1}+\cdots+\mathrm{a}_2\phi+\mathrm{a}_1 I=0\ , \tag{8.7}$$

where I denotes the identity tensor field, and $\phi^{k-1}(x),\ldots,\phi(x)$, I being linearly independent at all $x\in M$. The polynomial Q is called structural.

In order to give consistency to the examples we are going to describe, we refer the reader to [38] and [85], where the existence of polynomial structures on parallelizable or simply connected manifolds is proved, results that can be applied to show the existence of elements $u\in V=Hom(\mathbf{R}^n,\mathbf{R}^n)$, $u'\in V'=Hom(gl(n,\mathbf{R}),gl(n,\mathbf{R}))$ verifying equations similar to (8.7).

Thus, let $u\in V$, $u'\in V'$ be square matrices of ranks r and r', respectively, and satisfying the equations

$$Q(u)=u^k+\mathrm{a}_k u^{k-1}+\ldots+\mathrm{a}_2 u+\mathrm{a}_1 I=0\ , \tag{8.8}$$

$$Q'(u')=(u')^{k'}+\bar{\mathrm{a}}_{k'}(u')^{k'-1}+\ldots+\bar{\mathrm{a}}_2 u'+\bar{\mathrm{a}}_1 I=0\ , \tag{8.9}$$

$u^{k-1},\ldots,u$, I (resp. $(u')^{k'-1},\ldots,u'$, I) being linearly independent. Then $\tilde{u}=u+u'\in\widetilde{V}$ has rank $r+r'$ and satisfies the equation

$$\widetilde{Q}(\tilde{u})=\tilde{u}^{\tilde{k}}+\tilde{a}_{\tilde{k}}\tilde{u}^{\tilde{k}-1}+\ldots+\tilde{a}_2\tilde{u}+\tilde{a}_1 I=0\ , \tag{8.10}$$

where $\widetilde{Q}$ is the lowest common multiple of Q and Q', and $\tilde{u}^{\tilde{k}-1},\ldots,\tilde{u}$, I are linearly independent.

Therefore, if we consider a linear connection ω on M, the tensor field $\widetilde{J}$ on FM, given by (8.3), defines a polynomial structure on FM with structural polynomial $\widetilde{Q}$ and rank $r+r'$.

This provides a useful method for constructing polynomial G–structures on FM, a method that we shall exploit later on with some examples. Before that, let us remark that all the results in Section 8.2 apply to these polynomial structures, but they can be significantly reinforced if some supplementary hypothesis on $u' \in V'$ or on Q and Q' is taken.

More precisely, let us consider the following situation: assume that $u' \in V'$ is the homomorphism given by $u'(A) = A \cdot N$, taking $N \in gl(n, \mathbf{R})$ a fixed matrix. Then:

Lemma 8.3.2 *If rank $N = m$, then rank $u' = nm$.*

Proof. Let $N = (N_j^i)$ be the matrix expression of N with respect to the canonical basis of $\mathbf{R}^n$. Then, a direct computation proves that u' is given by

$$u'(E_i^j) = \sum_{k=1}^{n} N_k^i E_k^j \,,$$

and, therefore, u' is represented with respect to the natural basis of $gl(n, \mathbf{R})$ by the matrix

$$\begin{pmatrix} N_1^1 & \dots & 0 & N_1^2 & \dots & 0 & \dots & N_1^n & \dots & 0 \\ \vdots & \ddots & \vdots & \vdots & \ddots & \vdots & \ddots & \vdots & \ddots & \vdots \\ 0 & \dots & N_1^1 & 0 & \dots & N_1^2 & \dots & 0 & \dots & N_1^n \\ & \vdots & & \vdots & \vdots & \vdots & \vdots & & \vdots & \\ N_n^1 & \dots & 0 & \cdot & \dots & \cdot & \dots & N_n^n & \dots & 0 \\ \vdots & \ddots & \vdots & \vdots & \vdots & \vdots & \vdots & \vdots & \ddots & \vdots \\ 0 & \dots & N_n^1 & \cdot & \dots & \cdot & \dots & 0 & \dots & N_n^n \end{pmatrix} .$$

Then, we have:

Theorem 8.3.3 *Let*

(a): *$u \in V$ with rank $u = r$ and satisfying* (8.8);

(b): *$u' \in V'$ given by $u'(A) = N \cdot A$, for a fixed $N \in gl(n, \mathbf{R})$ with rank $N = m$, and N (and hence u') satisfying* (8.9).

Then:

(i): *$\tilde{u} = u + u' \in \tilde{V}$ has rank $(r + nm)$ and satisfies* (8.10).

(ii): *Let ω be a linear connection on M; then $\tilde{J}$ given by* (8.3) *defines on FM a polynomial structure of constant rank $r + nm$ and structural polynomial $\tilde{Q}$.*

(iii): $N_{\tilde{J}} = B \circ \widetilde{d\theta} + \lambda \circ \tilde{\Omega}$.

Proof. **(i)** and **(ii)** are obvious. To prove **(iii)**, it suffices to recall (8.5) and to check that, under the hypothesis for u', $\omega \widetilde{\wedge} \omega$ vanishes identically.□

Using this theorem we obtain an easy description of the examples considered by T. Okubo [73,74].

Example 1: $f(3,1)$–structure on FM

With the notation of Theorem 8.3.3, take $u = 0$ and

$$\begin{pmatrix} 0 & -I_m \\ I_m & 0 \end{pmatrix}$$

with $n = 2m$. Then *rank* $u' = n^2$ and the structural polynomials are

$$Q(u) = u = 0 \ , \ Q'(u') = (u')^2 + I \ .$$

Therefore $\tilde{u} = 0 + u'$ satisfies

$$\tilde{Q}(\tilde{u}) = \tilde{u}^3 + \tilde{u} = 0 \ ,$$

and *rank* $\tilde{u} = n^2$. Now, given the linear connection ω, $\tilde{J}$ is given by $\tilde{J} = \lambda \circ u' \circ \omega$, and since $\tilde{J}^3 + \tilde{J} = 0$ and *rank* $\tilde{J} = m^2$, it defines an $f(3,1)$–structure of *rank* n^2 on FM. Moreover

Theorem 8.3.4 *The $f(3,1)$–structure above is integrable if and only if ω is flat.*

Proof. It suffices to apply **(iii)** in Theorem 8.3.3 here by checking that $\widetilde{d\theta} = 0$ and $\tilde{\Omega} = 0$ for these particular choices of u and u'. Then, since $N_{\tilde{J}} = \lambda \circ \Omega$ the theorem is proved.□

In order to check that $\tilde{J}$ actually defines the same $f(3,1)$–structure that Okubo considered on FM, it suffices to write out the matrix representation of $\tilde{J}_p$ at an arbitrary point $p \in FM$ with respect to the canonical frame $\tilde{p}_0 = \{(Be)_p, (\lambda E)_p\}$, which is easy by direct computation.

We must remark that we are using an arbitrary linear connection ω on M, whereas Okubo's construction was based on a torsion free linear connection.

It is not difficult to reveal some properties of $\tilde{J}$. Consider the canonical projectors $\pi_1 = -\tilde{J}^2$, $\pi_2 = \tilde{J}^2 + I$. Then $T_1 = Im\ \pi_1$ is the vertical distribution on FM, and $T_2 = Im\ \pi_2$ is the horizontal distribution of ω. From [43] we know that

$$T_1 \text{ is integrable} \iff \pi_2 N_{\tilde{J}}(X,Y) = 0 \ ,$$

$$T_2 \text{ is integrable} \iff N_{\tilde{J}}(\pi_2 X, \pi_2 Y) = 0 \ .$$

Obviously, T_1 is always integrable and the integral manifolds are the fibres of FM on M, and T_2 is integrable if and only if ω is flat. Moreover, since Im $\tilde{J} = Im\ \pi_1$, it is easy to show that $\tilde{J}$ is always partially integrable [43].

Moreover, if we adapt to the present situation the terminology and definitions of the theory of f–structures, it is not difficult to prove that $\tilde{J}$ defines on FM

a framed $f(3,1)$–structure. In fact, $Be_1, \ldots, Be_n$ generate T_2, and the 1–forms $\theta^1, \ldots, \theta^n$ given by $\theta = \sum \theta^i \otimes e_i$ satisfy

$$\theta^i(Be_j) = \delta^i_j \ , \ \pi_2 = \sum Be_i \otimes \theta^i \ .$$

Nevertheless, the structure is not normal in general. For, in this case the necessary and sufficient conditions for normality are

$$\lambda \circ \Omega + B \circ d\theta = 0 \ , \ d\theta(X, \tilde{J}Y) + d\theta(\tilde{J}X, Y) = 0 \ ,$$

identities that are not satisfied even if ω is torsion free.

Remark 8.3.5 Note that in this example $\tilde{\alpha} = 0$ for every $\alpha \in \Lambda^2(FM, \mathbf{R}^n)$.

Example 2: $f(3,-1)$–structure on FM

For arbitrary n, take again $u = 0$ and

$$N = \begin{pmatrix} 0 & \ldots & 1 \\ \vdots & \ddots & \vdots \\ 1 & \ldots & 0 \end{pmatrix} .$$

Then $rank\ u' = n^2$, and the structural polynomials are

$$Q(u) = u = 0 \ , \ Q'(u') = (u')^2 - I = 0 \ .$$

Therefore $\tilde{u} = 0 + u'$ satisfies the equation

$$\tilde{Q}(\tilde{u}) = \tilde{u}^3 - \tilde{u} = 0 \ ,$$

and $rank\ \tilde{u} = n^2$.

If ω is a linear connection on M, then $\tilde{J}$ is given by $\tilde{J} = \lambda \circ u' \circ \omega$ and, since $\tilde{J}^3 - \tilde{J} = 0$ and $rank\ \tilde{J} = n^2$, it defines on FM an $f(3,-1)$–structure of $rank\ n^2$.

Theorem 8.3.4 on the integrability of $\tilde{J}$ is still valid in this case, as well as the other results of the previous example on partial integrability, etc. Finally, one easily checks that $\tilde{J}$ is, in fact, the $f(3,-1)$–structure on FM defined by Okubo, again without requiring the torsion free hypothesis for ω.

Example 3: $f(4,2)$–structure on FM

Assume $dim\ M = n = 2m$, and take $u \in V$ and $N \in gl(n, \mathbf{R})$ given by

$$u = \begin{pmatrix} 0 & 0 \\ I_m & 0 \end{pmatrix} , \ N = \begin{pmatrix} 0 & -I_m \\ I_m & 0 \end{pmatrix} .$$

Then $rank\ u = m$, $rank\ u' = n^2$, and the structural polynomials are

$$Q(u) = u^2 = 0 \ , \ Q'(u') = (u')^2 + I = 0 \ .$$

Hence $\tilde{u} = u + u'$ satisfies

$$\tilde{Q}(\tilde{u}) = \tilde{u}^4 + \tilde{u}^2 = 0 \ ,$$

and $rank\ \tilde{u} = 4m^2 + m$.

Now, for an arbitrary linear connection ω, $\tilde{J}$ given by (8.3) defines an $f(4,2)$–structure on FM of $rank$ $4m^2 + m$. The vanishing of $N_{\tilde{J}}$ is here also a necessary and sufficient condition for integrability, and **(iii)** in Theorem 8.3.3 gives its expression. Taking into account the results in [11], one easily obtains other properties of this structure. The projector operators π_1 and π_2 are given as in Example 1, and the distributions T_1 and T_2 are the same; with the terminology of [11], the structure is always c–partially integrable, and the partial integrability and the t–partial integrability are equivalent, and they hold if and only if $\Omega(\pi_2 X, \pi_2 Y) = 0$.

Example 4: $f(4,-2)$–structure on FM

With $n = 2m$, take $u \in V$ as in Example 3 and $N \in gl(n, \mathbf{R})$ as in Example 2. Then the structural polynomials are

$$Q(u) = u^2 = 0 \ , \ Q'(u') = (u')^2 - I = 0 \ ,$$

and for $\tilde{u} = u + u' \in \tilde{V}$,

$$\tilde{Q}(\tilde{u}) = \tilde{u}^4 - \tilde{u}^2 = 0 \ .$$

In this case $\tilde{J}$ defines on FM an $f(4,-2)$–structure of $rank$ $4m^2 + m$. If we now define $\pi_1 = \tilde{J}^2$, $\pi_2 = I - \tilde{J}^2$, results similar to those in Example 3 are still valid.

Example 5: A family of examples

Recall Theorem 8.3.3. Let $u \in V$ be a homomorphism of $rank$ k satisfying equation (8.8); then, the homomorphism $u' \in V'$ obtained by taking $N = u$

satisfies also (8.8) and has *rank* kn. Obviously, $Q = Q' = \tilde{Q}$ and $\tilde{J}$ defines on FM a polynomial structure of *rank* $k(n+1)$ and structural polynomial Q.

A direct computation proves the vanishing of the term $\widetilde{\omega \wedge \theta}$ in $\widetilde{d\theta}$ when $N = u$, and hence $N_{\tilde{J}} = B \circ \tilde{\Theta} + \lambda \circ \tilde{\Omega}$.

In fact, the almost complex structure on FM ($n = 2m$) constructed by Terrier [83] is an example in this family.

8.4 G-structures defined by (0,2)–tensor fields

We shall devote this section to the construction of Riemannian and almost symplectic structures on FM.

Let be

$$\tilde{V} = W^* \otimes W^* \quad , \quad W = \mathbf{R}^{n+n^2} \simeq \mathbf{R}^n \times gl(n,\mathbf{R}) \quad ,$$

and let $\tilde{\rho}: Gl(n+n^2,\mathbf{R}) \longrightarrow Gl(\tilde{V})$ be the canonical linear representation.

It is well known from the general theory of G–structures that there exists a one–to–one correspondence between $G_{\tilde{u}}$–structures on FM defined as $\tilde{t}^{-1}(\tilde{u})$, $\tilde{t}: FFM \longrightarrow \tilde{V}$ being a differentiable tensor of type $(\tilde{\rho}, \tilde{V})$ and $\tilde{V}_{\tilde{u}}$–valued, $G_{\tilde{u}}$ the isotropy group of $\tilde{u} \in \tilde{V}$, and tensor fields of type $(0,2)$ on FM defined by

$$\begin{array}{rrcl} & \tilde{\Phi}: T_pFM \times T_pFM & \longrightarrow & V_{\tilde{u}} \\ (8.11) & & & \\ & (X,Y) & \longmapsto & \tilde{t}(\tilde{p})(\tilde{p}^{-1}X, \tilde{p}^{-1}Y) \ , \end{array}$$

for all $\tilde{p} \in \pi^{-1}(p)$.

Among all the possible G–structures on FM defined by tensor fields of type $(0,2)$, we shall be interested in those that are defined from particular structures on the "model" spaces $\mathbf{R}^n$ and $gl(n,\mathbf{R})$ and ω–associated in the sense described in Section 8.1.

So let ω be a linear connection on M, and consider the vector spaces

$$V = (\mathbf{R}^n)^* \otimes (\mathbf{R}^n)^* \ , \ V' = gl(n,\mathbf{R}) \otimes gl(n,\mathbf{R})^* \ ,$$

and the canonical linear representations

$$\rho: Gl(n,\mathbf{R}) \longrightarrow Gl(V) \ , \ \rho': Gl(n^2,\mathbf{R}) \longrightarrow Gl(V') \ .$$

Next, if for $u \in V$ and $u' \in V'$ we put $\tilde{u} = u + u' \in \tilde{V}$, then

$$V_u \oplus V'_{u'} \subset \tilde{V}_{\tilde{u}} \ , \ j(G_u \times G_{u'}) \subset G_{\tilde{u}} \ ,$$

where j is the canonical injection.

From Definition 8.1.1 we know that the ω–associated $G_{\tilde{u}}$–structure on FM is defined by a differentiable tensor $\tilde{t}$ of type $(\tilde{\rho}, \tilde{V})$, and we can describe the associated tensor field $\tilde{\Phi}$ on FM as follows.

Theorem 8.4.1 *At every point $p \in FM$*

$$\tilde{\Phi}_p(X,Y) = u(\theta_p(X), \theta_p(Y)) + u'(\omega_p(X), \omega_p(Y)), \tag{8.12}$$

for all $X, Y \in T_pFM$.

Proof. It follows easily from (8.11) using the canonical frame $\tilde{p}_0 = \sigma_\omega(p)$, and taking into account the identities

$$\tilde{p}_0^{-1}(X) = \theta_p(X) + \omega_p(X) \ , \ \tilde{t}(\tilde{p}_0) = \tilde{u} = u + u' \ .\square$$

We shall now apply this construction to some particular cases.

<u>Almost symplectic structures on FM.</u>

We assume $dim\ M = n = 2m$.

Recall again that an $Sp(n, \mathbf{R})$–structure (almost symplectic structure) on M is uniquely determined by a differential 2–form of maximal rank; moreover, the structure is symplectic or integrable if and only if the almost symplectic form is closed.

Let both $u \in V$ and $u' \in V'$ be skew–symmetric and of maximal rank. Then, given a linear connection ω on M, $\tilde{\Phi}$ given by (8.12) is a differential 2–form and, moreover,

Theorem 8.4.2 *$\tilde{\Phi}$ defines on FM an almost symplectic structure, that is $\tilde{\Phi}$ has maximal rank.*

Proof. Choose $\{\bar{e}_i\}$ and $\{\bar{E}_i^j\}$ bases of $\mathbf{R}^n$ and $gl(n, \mathbf{R})$ adapted to u and u', respectively; that is, if $\{\alpha^i\}$ and $\{\alpha_j^i\}$ denote their dual bases, then

$$u = \sum_{i=1}^{n} \alpha^i \wedge \alpha^{i+m} \ , \ u' = \sum_{j=1}^{n} \sum_{i=1}^{2m^2} \alpha_j^i \wedge \alpha_j^{i+2m^2} \ .$$

Thus, if we put

$$\theta = \sum_{i=1}^{n} \theta^i \otimes \bar{e}_i \ , \ \omega = \sum_{i,j=1}^{n} \omega_j^i \otimes \bar{E}_i^j \ ,$$

we obtain

$$\tilde{\phi} = \sum_{i=1}^{n} \theta^i \wedge \theta^{i+m} + \sum_{j=1}^{n} \sum_{i=1}^{2m^2} \omega_j^i \wedge \omega_j^{i+2m^2} \ .\square$$

Remark 8.4.3 In general, for arbitrary $u \in V$ and $u' \in V'$ with $rank\ u = r$, $rank\ u' = r'$, one has $rank\ \tilde{\Phi} = r + r'$.

In order to characterize the integrability of these almost symplectic structures on FM, we introduce a new definition.

Definition 8.4.4 *Let $u \in F$ and $u' \in F'$ and take ω a linear connection on M. Then:*

(i): *if $\alpha \in \Lambda^2(FM, \mathbf{R}^n)$, $\tilde{\alpha} \in \Lambda^3(FM, \mathbf{R})$ is the differential 3-form on FM given by*

$$3\tilde{\alpha}(X,Y,Z) = u(\alpha(X,Y), \theta Z) + u(\alpha(Y,Z), \theta X) + u(\alpha(Z,X), \theta Y) \; ;$$

(ii): *if $\alpha \in \Lambda^2(FM, gl(n, \mathbf{R}))$, $\tilde{\alpha} \in \Lambda^3(FM, \mathbf{R})$ is the differential 3-form on FM given by*

$$3\tilde{\alpha}(X,Y,Z) = u'(\alpha(X,Y), \omega Z) + u'(\alpha(Y,Z), \omega X) + u'(\alpha(Z,X), \omega Y) \; ,$$

X, Y, Z being arbitrary vector fields on FM.

Theorem 8.4.5 $d\tilde{\Phi} = 2\widetilde{d\theta} + 2\widetilde{d\omega}$.

Proof. It suffices to check the identity in the following cases:

(a): $X = B_1$, $Y = B_2$, $Z = B_3$, $B_i = B\xi_i$, $\xi_i \in \mathbf{R}^n$, $i = 1,2,3$, and linearly independent;

(b): $X = B_1$, $Y = B_2$, $Z = \lambda A$, $B_i = B\xi_i$, $\xi_i \in \mathbf{R}^n$, $A \in gl(n, \mathbf{R})$, being $A \neq 0$ and ξ_1, ξ_2 linearly independent;

(c): $X = B\xi$, $Y = \lambda A_1$, $Z = \lambda A_2$, $\xi \in \mathbf{R}^n$, $A_i \in gl(n, \mathbf{R})$, $\xi \neq 0$, A_1 and A_2 linearly independent;

(d): $X = \lambda A_1$, $Y = \lambda A_2$, $Z = \lambda A_3$, $A_i \in gl(n, \mathbf{R})$, $i = 1,2,3$, and linearly independent.□

In the usual terminology of symplectic manifold theory, it follows:

Proposition 8.4.6 *Let H_p and V_p be the horizontal and vertical subspaces at $p \in FM$. Then, with respect to the almost symplectic structure defined by $\tilde{\Phi}$ on FM,*

(1): *H_p and V_p are $\tilde{\Phi}$-orthogonal symplectic subspaces of T_pFM;*

(2): *the fibres are almost symplectic submanifolds of FM;*

(3): *if ω is flat, then the integral manifolds of the horizontal distribution are almost symplectic submanifolds.*

Proof. **(1)** is a direct consequence of (8.12); **(2)** and **(3)** follow from **(1)**.□

Next, assume that there exists an almost symplectic form Φ_M on M; that is Φ_M is a differential 2-form on M of maximal rank such that the subbundle P of adapted frames is given by

$$P = \{p \in FM \,|\, (\Phi_M)_{\pi(p)}(p\xi, p\xi') = u(\xi, \xi') \; , \; \xi, \xi' \in \mathbf{R}^n\} \; ;$$

this means that Φ_M is "modeled" by $u \in V$. Then,

Theorem 8.4.7 *On P*

$$\tilde{\Phi}(X^H, Y^H) = (\Phi_M(X,Y) \circ \pi) \ ,$$

for all X, Y vector fields on M.

Proof. Let $p \in P$; from (8.12), $\tilde{\Phi}_p(X^H, Y^H) = u(\theta_p(X^H), \theta_p(Y^H))$. But

$$\theta_p(X^H) = p^{-1}(X_{\pi(p)}) \ , \ \theta_p(Y^H) = p^{-1}(Y_{\pi(p)})$$

and hence

$$\begin{aligned} \tilde{\Phi}_p(X^H, Y^H) &= u(p^{-1}X_{\pi(p)}, p^{-1}Y_{\pi(p)}) \\ &= (\Phi_M)_{\pi(p)}(X_{\pi(p)}, Y_{\pi(p)}) \\ &= (\Phi_M(X,Y))(\pi(p)) \ .\square \end{aligned}$$

Theorem 8.4.8 *With the same hypothesis as above and ω an adapted connection on P, then*

$$2\tilde{\Theta}(X^H, Y^H, Z^H) = (d\Phi_M(X,Y,Z)) \circ \pi \ ,$$

on P for all vector fields X, Y, Z on M.

Proof. On the one hand, we have

$$\begin{aligned} 3d\tilde{\Phi}(X^H, Y^H, Z^H) =& X^H(\tilde{\Phi}(Y^H, Z^H)) + Y^H(\tilde{\Phi}(Z^H, X^H)) \\ &+ Z^H(\tilde{\Phi}(X^H, Y^H)) - \tilde{\Phi}([X^H, Y^H], Z^H) \\ &- \tilde{\Phi}([Y^H, Z^H], X^H) - \tilde{\Phi}([Z^H, X^H], Y^H) \ . \end{aligned}$$

Using Theorem 8.4.1, on P we have

$$X^H(\tilde{\Phi}(Y^H, Z^H)) = (X(\Phi_M(Y,Z)) \circ \pi \ ,$$

and, if h denotes the horizontal projector associated to ω, since ω is adapted,

$$\begin{aligned} \tilde{\Phi}([X^H, Y^H], Z^H) &= \tilde{\Phi}(h[X^H, Y^H], Z^H) \\ &= \tilde{\Phi}([X,Y]^H, Z^H) \\ &= (\Phi_M([X,Y], Z)) \circ \pi \ , \end{aligned}$$

on P; therefore,

$$d\tilde{\Phi}(X^H, Y^H, Z^H) = (d\Phi_M(X,Y,Z)) \circ \pi \ .$$

On the other hand,

$$d\tilde{\Phi} = 2\widetilde{d\theta} + 2\widetilde{d\omega}$$

(see Theorem 8.4.5); but

$$d\theta(X^H, Y^H) = \Theta(X^H, Y^H) ,$$

so

$$\widetilde{d\theta}(X^H, Y^H, Z^H) = \tilde{\Theta}(X^H, Y^H, Z^H)$$

and the result follows because $\widetilde{d\omega}(X^H, Y^H, Z^H) = 0$.□

This theorem has two interesting applications, both being well known results of the theory of symplectic structures.

Corollary 8.4.9 *If there exists on the almost symplectic manifold a torsion free connection leaving Φ_M parallel, then the structure is symplectic, that is integrable.*

Proof. It suffices to prove that $d\Phi_M = 0$. But $\Theta = 0$ implies $\tilde{\Theta} = 0$, and by Theorem 8.4.8, $d\Phi_M = 0$.□

Corollary 8.4.10 *If ω is a linear connection on M leaving the almost symplectic form Φ_M parallel, and if T denotes the torsion tensor of ω, then*

$$3d\Phi_M(X, Y, Z) = \sum_{X,Y,Z} \Phi_M(T(X, Y), Z) \tag{8.13}$$

for all vector fields X, Y, Z on M ($\sum_{X,Y,Z}$ means cyclic sum).

Proof. It is similar to that of Corollary 8.2.8.□

Remark 8.4.11 Identity (8.13) justifies Definition 8.4.4.

<u>Riemannian structures on FM.</u>

Assume $u \in V$ and $u' \in V'$ both being symmetric and positive definite, that is defining inner products on $\mathbf{R}^n$ and $gl(n, \mathbf{R})$, respectively. Then, given a linear connection ω on M, the associated tensor field $\tilde{\Phi}$ on FM, determined by (8.12), is also symmetric and positive definite. Hence

Theorem 8.4.12 *$\tilde{\Phi}$ is a Riemannian metric on FM.*

Proof. It suffices to prove that $\tilde{\Phi}$ is positive definite, because the rest of the conditions are obvious. For all $X \in T_pFM$,

$$\tilde{\Phi}_p(X, X) = u(\theta_p(X), \theta_p(X)) + u'(\omega_p(X), \omega_p(X)) \geq 0 ,$$

and the equality holds only when

$$u(\theta_p(X), \theta_p(X) = u'(\omega_p(X), \omega_p(X)) = 0 \ .$$

But u and u' both are positive definite, so $\tilde{\Phi}_p(X, X) = 0$ if and only if $\theta_p(X) = \omega_p(X) = 0$, that is $X = 0$.□

Observe that $\tilde{\Phi}$ makes the horizontal and the vertical distribution on FM mutually orthogonal.

Let us now consider on FM the canonical flat connection induced by the absolute parallelism associated to ω (see Section 6.2); its covariant derivative $\overline{\nabla}$ is given as follows: for any vector field

$$Y = \sum_{i=1}^{n} \mathrm{a}^i Be_i + \sum_{i,j=1}^{n} \mathrm{b}_i^j \lambda E_j^i \ ,$$

on FM,

$$\overline{\nabla}_X Y = \sum_{i=1}^{n} (X\mathrm{a}^i) Be_i + \sum_{i,j=1}^{n} (X\mathrm{b}_i^j) \lambda E_j^i \ .$$

Then,

Proposition 8.4.13 $\overline{\nabla}$ *is a metric connection, that is* $\overline{\nabla}\tilde{\Phi} = 0$.

Proof. It suffices to check the identity

$$X(\tilde{\Phi}(Y, Z)) = \tilde{\Phi}(\overline{\nabla}_X Y, Z) + \tilde{\Phi}(Y, \overline{\nabla}_X Z)$$

for arbitrary vector fields X, Y, Z on FM. The result follows easily from the identities

$$\begin{aligned} \tilde{\Phi}(Be_i, Be_j) &= u(e_i, e_j) \ , \\ \tilde{\Phi}(Be_i, \lambda E_k^j) &= 0 \ , \\ \tilde{\Phi}(\lambda E_j^i, \lambda E_i^k) &= u'(E_j^i, E_i^k) \ . \end{aligned}$$ □

Moreover, if $\widetilde{\nabla}$ denotes the Levi–Civita connection of $\tilde{\Phi}$, then we obtain

Corollary 8.4.14 *The connections* $\overline{\nabla}$ *and* $\widetilde{\nabla}$ *are always different.*

Proof. It is a consequence of Corollary 6.2.31.□

Next, assume that there is given on M a Riemannian metric Φ_M "modeled" on $u \in V$, that is such that the bundle OM of orthonormal frames is given by:

$$OM = \{p \in FM \mid (\Phi_M)_{\pi(p)}(p\xi, p\xi') = u(\xi, \xi'), \xi, \xi' \in \mathbf{R}^n\} \ .$$

Then,

Theorem 8.4.15 *On OM*

$$\tilde{\Phi}(X^H, Y^H) = (\Phi_M(X,Y)) \circ \pi$$

for all vector fields X, Y on M.

Proof. Similar to that of Theorem 8.4.7.□

Theorem 8.4.16 *If ω is the Levi-Civita connection of Φ_M and its curvature vanishes, then on OM*

$$\widetilde{\nabla}_{X^H} Y^H = (\nabla_X Y)^H$$

for all vector fields X, Y on M.

Proof. Recall that

$$\begin{aligned} 2\tilde{\Phi}(\widetilde{\nabla}_{X^H} Y^H, \tilde{Z}) &= X^H \tilde{\Phi}(Y^H, \tilde{Z}) + Y^H \tilde{\Phi}(X^H, \tilde{Z}) \\ &\quad - \tilde{Z}\tilde{\Phi}(X^H, Y^H) + \tilde{\Phi}([X^H, Y^H], \tilde{Z}) \\ &\quad + \tilde{\Phi}([\tilde{Z}, X^H], Y^H) + \tilde{\Phi}(X^H, [\tilde{Z}, Y^H]) \ , \end{aligned}$$

$\tilde{Z}$ being an arbitrary vector field on OM. Then, using the previous theorem, for any vector field Z on M,

$$\begin{aligned} 2\tilde{\Phi}(\widetilde{\nabla}_{X^H} Y^H, Z^H) &= X^H(\Phi_M(X,Z) \circ \pi) + Y^H(\Phi_M(X,Z) \circ \pi) \\ &\quad - Z^H(\Phi_M(X,Y) \circ \pi) + \tilde{\Phi}([X,Y]^H, Z^H) \\ &\quad + \tilde{\Phi}([Z,X]^H, Y^H) + \tilde{\Phi}(X^H, [Z,Y]^H) \end{aligned}$$

on OM. Since $\Omega = 0$ implies $[X^H, Y^H] = [X,Y]^H$, then

$$2\tilde{\Phi}(\widetilde{\nabla}_{X^H} Y^H, Z^H) = (\Phi_M(\nabla_X Y, Z)) \circ \pi \ . \tag{8.14}$$

On the other hand, since $[\lambda A, X^H] = 0$ for any $A \in \mathfrak{o}(n, \mathbf{R})$, the Lie algebra of the orthogonal group, and because the horizontal and the vertical distributions are orthogonal, then

$$2\tilde{\Phi}(\widetilde{\nabla}_{X^H} Y^H, \lambda A) = -\lambda A(\Phi_M(X,Y) \circ \pi) = 0 \tag{8.15}$$

on OM. The result follows combining (8.14) and (8.15).□

Corollary 8.4.17 *With the hypothesis of* Theorem 8.4.16, *the horizontal lifts to OM of geodesics of ∇ are geodesics of $\widetilde{\nabla}$.*

To end this section we return to the Examples 1 and 2.

Let $\tilde{J}$ be the $f(3,1)$–structure on FM of Example 1. A Riemannian metric $\tilde{G}$ on FM is called adapted to $\tilde{J}$ if

$$\tilde{G}(X, \tilde{J}Y) = -\tilde{G}(\tilde{J}X, Y) .$$

Then, a direct computation from the definitions involved leads to the following result: let $\tilde{\Phi}$ be the Riemannian metric on FM given by (8.9) and constructed with respect to the same connection ω as $\tilde{J}$; then $\tilde{\Phi}$ is adapted to $\tilde{J}$ if and only if u' is Hermitian with respect to the almost complex structure used on $gl(n, \mathbf{R})$ in constructing $\tilde{J}$. This means that on FM there always exist metrics compatible with $\tilde{J}$.

A similar result follows in the situation considered in Example 2, changing the condition of being an Hermitian metric to that of being compatible with the structure considered on $gl(n, \mathbf{R})$ in this case.

8.5 Applications to almost complex and Hermitian structures

Throughout this Section we shall assume $n = 2m + 1$.

Let us consider $\mathbf{R}^n$ endowed with its trivial Abelian Lie algebra structure, and $gl(n, \mathbf{R})$ with its natural Lie algebra structure. Then, the canonical representation

$$gl(n, \mathbf{R}) \longrightarrow Der(\mathbf{R}^n)$$

allows us to introduce a Lie algebra structure on the vector space $\mathbf{R}^n \times gl(n, \mathbf{R})$ by defining the bracket product as follows:

$$[(\alpha_1, A_1), (\alpha_2, A_2)] = (A_1(\alpha_2) - A_2(\alpha_1), [A_1, A_2])$$

for all $\alpha_1, \alpha_2 \in \mathbf{R}^n$ and $A_1, A_2 \in gl(n, \mathbf{R})$. It is well known that, with this structure, $\mathbf{R}^n \times gl(n, \mathbf{R})$ is canonically isomorphic to the Lie algebra of the affine real group $A(n, \mathbf{R})$.

Let (u, η, ξ) (resp. (u', η', ξ')) be a triple with $u \in Hom(\mathbf{R}^n, \mathbf{R}^n)$, $\eta \in (\mathbf{R}^n)^*$ and $\xi \in \mathbf{R}^n$ (resp. $u' \in Hom(gl(n, \mathbf{R}), gl(n, \mathbf{R}))$, $\eta' \in gl(n, \mathbf{R})^*$ and $\xi' \in gl(n, \mathbf{R})$), satisfying the following identities:

$$u^2 = -I + \eta \otimes \xi \quad , \ \eta(\xi) = 1 .$$

$$(\text{res. } (u')^2 = -I + \eta' \otimes \xi' \quad , \ \eta'(\xi') = 1 .)$$

Then,

$$u(\xi) = 0 \quad , \eta \circ u = 0 \; , \qquad rank\ u = 2m \; ,$$
$$u'(\xi') = 0 \quad , \eta' \circ u' = 0 \; , \quad rank\ u' = n^2 - 1 \; .$$

Such a triple (u, η, ξ) (resp. (u', η', ξ')) will be called an almost contact structure on $\mathbf{R}^n$ (resp. on $gl(n, \mathbf{R})$).

Let (u, η, ξ) and (u', η', ξ') be almost contact structures on $\mathbf{R}^n$ and $gl(n, \mathbf{R})$, and define a homomorphism

$$\tilde{u}: \mathbf{R}^n \times gl(n, \mathbf{R}) \longrightarrow \mathbf{R}^n \times gl(n, \mathbf{R})$$

by setting

$$\tilde{u}(\alpha, A) = (u(\alpha) + \eta'(A) \cdot \xi, u'(A) - \eta(\alpha) \cdot \xi') \; , \; \alpha \in \mathbf{R}^n, A \in gl(n, \mathbf{R}) \; .$$

A direct computation shows that $\tilde{u}^2 = -I$, hence $\tilde{u}$ defines an almost complex structure on $\mathbf{R}^n \times gl(n, \mathbf{R})$. Taking into account the Lie algebra structure of $\mathbf{R}^n \times gl(n, \mathbf{R})$, we can define a sort of Nijenhuis tensor of $\tilde{u}$ by:

$$\begin{aligned} N_{\tilde{u}}((\alpha_1, A_1), (\alpha_2, A_2)) = & [\tilde{u}(\alpha_1, A_1), \tilde{u}(\alpha_2, A_2)] - \tilde{u}[\tilde{u}(\alpha_1, A_1), (\alpha_2, A_2)] \\ & - \tilde{u}[(\alpha_1, A_1), \tilde{u}(\alpha_2, A_2)] - [(\alpha_1, A_1), (\alpha_2, A_2)] \end{aligned}$$

for all $(\alpha_i, A_i) \in \mathbf{R}^n \times gl(n, \mathbf{R})$.

At this point, and before proceeding to the study on FM, we must note the following facts:

Remark 8.5.1 Our definitions have been inspired by those of Morimoto in [66]; in his article, Morimoto only considers normal almost contact structures, but it is not difficult to recover his definitions by imposing the appropriate conditions of normality. A more general algebraic model for the situation we are considering has been studied by Oubiña [75].

Remark 8.5.2 The vanishing of $N_{\tilde{u}}$ cannot be easily guaranteed, as it follows from Oubiña's results. For instance, if one considers on $\mathbf{R}^n$ the standard contact structure, and the one it induces on $\mathbf{R}^n \times gl(n, \mathbf{R})$, a direct computation shows that $N_{\tilde{u}}$ does not vanish even in this simple case.

Nevertheless, it is not difficult to describe an example with vanishing Nijenhuis torsion.

<u>An Example.</u>

Consider on $\mathbf{R}^n$ the following triple (u, η, ξ); for any point

$$\alpha = (\alpha^1, \alpha^2, \ldots, \alpha^{2m}, \alpha^{2m+1}) \in \mathbf{R}^n$$

define

$$u(\alpha) = (-\alpha^2, \alpha^1, -\alpha^4, \alpha^3, \ldots, -\alpha^{2m}, \alpha^{2m-1}, 0),$$

and take $\eta \in (\mathbf{R}^n)^*$ as the dual form of $\xi = (0, \ldots, 0, 1) \in \mathbf{R}^n$.

On $gl(n, \mathbf{R})$ we take $\eta' \in gl(n, \mathbf{R})^*$ as the dual form of $\xi' = E_n^n \in gl(n, \mathbf{R})$, and define u' as follows: for $A = (A_j^i) \in gl(n, \mathbf{R})$,

$$u'(A) = \begin{pmatrix} -A_1^2 & A_1^1 & & -A_1^{2m} & A_1^{2m-1} & -A_2^{2m+1} \\ -A_2^2 & A_2^1 & \cdot & -A_2^{2m} & A_2^{2m-1} & A_1^{2m+1} \\ \vdots & \vdots & \vdots & \vdots & \vdots & \vdots \\ \cdot & \cdot & \cdot & \cdot & \cdot & -A_{2m}^{2m+1} \\ -A_{2m}^2 & A_{2m}^1 & \cdot & -A_{2m}^{2m} & A_{2m}^{2m-1} & A_{2m-1}^{2m+1} \\ -A_{2m+1}^2 & A_{2m+1}^2 & \cdot & A_{2m+1}^{2m} & A_{2m+1}^{2m-1} & 0 \end{pmatrix} .$$

Now, using the results in [75], or by direct computation, one checks the vanishing of $N_{\tilde{u}}$. (We thank J.A. Oubiña for communicating to us this example.)

Let be M a differentiable manifold, $dim\ M = n = 2m + 1$, and let ω be a linear connection on M. Assume given almost contact structures $(u, \eta, \xi,)$ and (u', η', ξ') on $\mathbf{R}^n$ and $gl(n, \mathbf{R})$, respectively. Then we define a tensor field $\widetilde{J}$ on FM of type $(1,1)$ by

$$\widetilde{J} = B \circ u \circ \theta + (\eta' \circ \omega) \otimes B\xi + \lambda \circ u' \circ \omega - (\eta \circ \theta) \otimes \lambda\xi' ,$$

that is,

$$\widetilde{J}_p(X) = (Bu(\theta_p(X))_p + \eta'(\omega_p(X))(B\xi)_p + (\lambda u'(\omega_p(X))_p - (\theta_p(X))(\lambda\xi')_p$$

for all $X \in T_pFM$ and any $p \in FM$.

Theorem 8.5.3 *$\widetilde{J}$ defines an almost complex structure on FM.*

Proof. Direct computation.□

In order to determine $N_{\widetilde{J}}$ we introduce the following definition:

Definition 8.5.4 *Let $\tilde{u}$ be the almost complex structure on $\mathbf{R}^n \times gl(n, \mathbf{R})$ associated to (u, η, ξ) and (u', η', ξ'). If Φ is a differential 2-form on FM with values in $\mathbf{R}^n$ or in $gl(n, \mathbf{R})$, we shall consider it as $\mathbf{R}^n \times gl(n, \mathbf{R})$-valued in the natural way; then, we define $\widetilde{\Phi} \in \Lambda^2(FM, \mathbf{R}^n \times gl(n, \mathbf{R}))$ by setting*

$$\widetilde{\Phi}(X, Y) = 2\{-\Phi(\widetilde{J}X, \widetilde{J}Y) + \tilde{u}\Phi(\widetilde{J}X, Y) + \tilde{u}\Phi(X, \widetilde{J}Y) + \Phi(X, Y)\} .$$

Theorem 8.5.5 *For all vector fields X, Y on FM,*

$$N_{\tilde{J}}(X,Y) = (B+\lambda)\circ(\widetilde{d\theta}+\widetilde{d\omega})(X,Y) ,$$

where B (resp. λ) acts on the $\mathbf{R}^n$-component (resp. on the $gl(n,\mathbf{R})$-component) of the $\mathbf{R}^n \times gl(n,\mathbf{R})$-valued 2-form $\widetilde{d\theta}+\widetilde{d\omega}$.

Proof. It suffices to compute the identity in the three basic cases, as in the proof of Theorem 8.2.4, and use the identities

$$\widetilde{d\theta} = -\widetilde{\omega\wedge\theta}+\tilde{\Theta} \quad , \quad \widetilde{d\omega} = -\widetilde{\omega\wedge\omega}+\tilde{\Omega} .\square$$

In order to show the influence of the torsion Θ and the curvature Ω of ω, and also of $N_{\tilde{u}'}$ on the integrability of $\tilde{J}$, we need a preliminary lemma.

Lemma 8.5.6 *For all vector fields X, Y on FM*

$$(\widetilde{\omega\wedge\theta}+\widetilde{\omega\wedge\omega})(X,Y) = -N_{\tilde{u}'}((\theta(X),\omega(X)),(\theta(Y),\omega(Y))).$$

Proof. Again the proof is routine, checking the identity in the three basic cases.□

Theorem 8.5.7

$$N_{\tilde{J}} = (B+\lambda)(\tilde{\Theta}+\tilde{\Omega}+N_{\tilde{u}}\circ(\theta,\omega)) ,$$

where

$$(N_{\tilde{u}}\circ(\theta,\omega))(X,Y) = N_{\tilde{u}}((\theta(X),\omega(X)),(\theta(Y),\omega(Y))) .$$

Proof. Combine Theorem 8.5.5 and Lemma 8.5.6.□

Obviously, $N_{\tilde{u}}\circ(\theta,\omega)$ vanishes if and only if $N_{\tilde{u}}$ vanishes; therefore

Corollary 8.5.8 *If the almost complex structure $\tilde{u}$ on $\mathbf{R}^n \times gl(n,\mathbf{R})$ has zero Nijenhuis torsion, then*

$$N_{\tilde{J}} = (B+\lambda)(\tilde{\Theta}+\tilde{\Omega}) .$$

Observe that this Corollary applies, in particular, when the almost contact structures (u,η,ξ) and (u',η',ξ') are those defined in the Example described above.

If we now consider on M another linear connection ω', and the same almost contact structures on $\mathbf{R}^n$ and $gl(n,\mathbf{R})$, then the associated almost complex structure $\tilde{J}'$ on FM is related to $\tilde{J}$ by

$$\tilde{J}'-\tilde{J} = (\eta'\circ\mu)\otimes B\xi - \lambda\circ\mu\circ B\circ u\circ\theta + \lambda\circ u'\circ\mu - (\eta'\circ\omega')\otimes\lambda(\mu(B\xi)) ,$$

with $\mu = \omega-\omega'$.

Remark 8.5.9 This construction of $\tilde{J}$ can be generalized, following the ideas of Tanno [82] as follows: let ω, (u, η, ξ) and (u', η', ξ') be as before, and let (a,b) be a pair of real numbers with $\mathrm{a} \neq 0$. Then, define

$$\tilde{u} \simeq \tilde{u}(\mathrm{a}, \mathrm{b}) \colon \mathbf{R}^n \times gl(n, \mathbf{R}) \longrightarrow \mathbf{R}^n \times gl(n, \mathbf{R})$$

by

$$\begin{aligned} \tilde{u}(\alpha, A) &= (u(\alpha) + (\mathrm{b}\eta(\alpha) + \mathrm{a}\eta'(A)) \cdot \xi, \\ &\qquad u'(A) - (\mathrm{b}\eta'(A) + \mathrm{a}^{-1}(1 + \mathrm{b}^2)\eta(\alpha)) \cdot \xi') \end{aligned}$$

and $\tilde{u}^2 = -I$. Analogously, there is defined a tensor field $\tilde{J} \equiv \tilde{J}(\mathrm{a}, \mathrm{b})$ of type $(1,1)$ on FM by

$$\begin{aligned} \tilde{J} = {} & B \circ u\theta + (\mathrm{a}(\eta' \circ \omega) + \mathrm{b}(\eta \circ \theta)) \otimes B\xi \\ & + \lambda \circ u' \circ \omega - (\mathrm{b}(\eta' \circ \omega) + \mathrm{a}^{-1}(1 + \mathrm{b}^2)(\eta \circ \theta)) \otimes \lambda\xi' , \end{aligned}$$

and a direct computation shows that $\tilde{J}^2 = -I$. The structures considered before are just $\tilde{u}(1,0)$ and $\tilde{J}(1,0)$. Moreover, simple computations show easily that Theorems 8.5.5 and 8.5.7 and Corollary 8.5.8 are still valid in this general case.

Let us return again to the previous situation (that is, $\mathrm{a} = 1$, $\mathrm{b} = 0$), and assume that M possesses an almost contact structure defined by the triple $(\bar{\phi}, \bar{\eta}, \bar{\xi})$, $\bar{\phi}$ being a tensor field of type $(1,1)$, $\bar{\eta}$ a 1-form and $\bar{\xi}$ a vector field satisfying

$$\bar{\phi}^2 = -I + \bar{\eta} \otimes \bar{\xi} \quad , \quad \bar{\eta}(\bar{\xi}) = 1 \ .$$

Furthermore, assume that the almost contact structure on M is "modeled" on the almost contact structure (u, η, ξ) on $\mathbf{R}^n$, namely the subbundle $P \subset FM$ of adapted frames consists of points $p \in FM$ such that

$$\bar{\phi}_{\pi(p)}(X) = p \circ u \circ p^{-1}(X) \quad , \quad \bar{\xi}_{\pi(p)} = p(\xi) \quad , \quad \bar{\eta}_{\pi(p)}(X) = \eta(p^{-1}(X)) \ ,$$

for all $X \in T_{\pi(p)}M$.

Now, if $\tilde{J}$ is the almost complex structure on FM defined by $\tilde{J}$ above, then for any vector field X on M

$$(\tilde{J}X^H)_p = (Bu(\theta_p(X^H)))_p - \eta(\theta_p(X^H)) \cdot (\lambda\xi')_p \ .$$

Since $Y = (Bu(\theta_p(X^H)))_p \in T_pFM$ is the unique tangent vector satisfying

$$\pi(Y) = p \circ u \circ \theta_p(X^H) = p \circ u \circ p^{-1}(X_{\pi(p)}) \ ,$$

and $\{\bar{\phi}X\}^H_p = Z$ is the unique horizontal tangent vector at p such that $\pi(Z) =$

$\phi(X_{\pi(p)})$, then $Y = \{\phi X\}^H_p$ if $p \in P$.

Also, if $p \in P$ one has

$$\eta(\theta_p(X^H)) = \eta(p^{-1}(X_{\pi(p)})) = \bar{\eta}_{\pi(p)}(X_{\pi(p)}) ,$$

and the following theorem is proved.

Theorem 8.5.10 *On P*

$$\tilde{J}(X^H) = (\bar{\phi}X)^H - \bar{\eta}(X) \cdot \lambda\xi' , \tag{8.16}$$

for all vector field X on M.

Theorem 8.5.11 *If ω is adapted to the almost contact structure on M, then on P*

$$B\tilde{\Theta}(X^H, Y^H) = \{N_{\bar{\phi}} + 2d\bar{\eta} \otimes \bar{\xi})(X, Y)\}^H - B \circ (N_{\tilde{u}} \circ (\theta, \omega))(X^H, Y^H)) ,$$

where $N_{\bar{\phi}}$ denotes the Nijenhuis torsion of $\bar{\phi}$, and X, Y are arbitrary vector fields on M.

Proof. Firstly,

$$hN_{\tilde{J}}(X^H, Y^H) = B \circ (\tilde{\Theta} + \tilde{\Omega} + N_{\tilde{u}} \circ (\theta, \omega))(X^H, Y^H) .$$

Next, from Theorem 8.5.10 and since $h[X^H, Y^H] = [X, Y]^H$, it follows

$$\begin{aligned} &h\,N_{\tilde{J}}(X^H, Y^H) \\ &= [\bar{\phi}X, \bar{\phi}Y]^H - h\{\bar{\eta}(Y)[(\bar{\phi}X)^H, \lambda\xi'] + (\bar{\phi}X)^H(\bar{\eta}(Y)) \cdot \lambda\xi'\} \\ &-h\{\bar{\eta}(X)[\lambda\xi', (\bar{\phi}Y)^H] - (\bar{\phi}Y)^H(\bar{\eta}(X)) \cdot \lambda\xi'\} - h\tilde{J}[(\bar{\phi}X)^H, Y^H] \\ &+h\tilde{J}\{\bar{\eta}(X)[\lambda\xi', Y^H] - Y^H(\bar{\eta}(X)) \cdot \lambda\xi'\} - h\tilde{J}[X^H, (\bar{\phi}Y)^H] \\ &+h\tilde{J}\{\bar{\eta}(Y)[X^H, \lambda\xi'] + X^H(\bar{\eta}(Y)) \cdot \lambda\xi'\} - [X, Y]^H . \end{aligned}$$

Now, for any vector field Z on FM

$$h\tilde{J}(Z) = \tilde{J}(hZ) + \eta'(\omega(Z)) \cdot B\xi + \eta(\theta(hZ)) \cdot \lambda\xi' ,$$

and a direct computation leads to

$$h\tilde{J}[(\bar{\phi}X)^H, Y^H] = Bu\theta([\bar{\phi}X, Y]^H) + \eta'(\omega[(\bar{\phi}X)^H, Y^H]) \cdot B\xi .$$

Next, on P

$$\begin{aligned}\tilde{J}[\bar{\phi}X, Y]^H &= Bu\theta([\bar{\phi}X, Y]^H) - \eta(\theta[\bar{\phi}X, Y]^H) \cdot \lambda\xi' \\ &= \{\bar{\phi}[\bar{\phi}X, Y]\}^H - \eta([\bar{\phi}X, Y]) \cdot \lambda\xi' ,\end{aligned}$$

and taking into account the structure equation of ω, on P we obtain

$$h\tilde{J}[(\bar{\phi}X)^H, Y^H] = \{\bar{\phi}[\bar{\phi}X, Y]\}^H - 2\eta'(\Omega((\bar{\phi}X)^H, Y^H)) \cdot B\xi .$$

Hence, since $h\tilde{J}(\lambda\xi') = B\xi$,

$$\begin{aligned}h\, N_{\tilde{J}}(X^H, Y^H) &= [\bar{\phi}X, \bar{\phi}Y]^H - \{\bar{\phi}[\bar{\phi}X, Y]\}^H \\ &-\{\bar{\phi}[X, \bar{\phi}Y]\}^H - [X, Y]^H + 2\eta'(\Omega((\bar{\phi}X)^H, Y^H)) \cdot B\xi \\ &-Y^H(\bar{\eta}(X)) \cdot B\xi + 2\eta'(\Omega(X^H, (\bar{\phi}Y)^H)) \cdot B\xi \\ &+X^H(\bar{\eta}(Y)) \cdot B\xi .\end{aligned}$$

Now, $Y^H(\bar{\eta}(X)) = Y(\bar{\eta}(X)) \circ \pi$, and the fact that ω is adapted to the structure implies that $B\xi = \xi^H$ on P; them from the definition of $\tilde{\Omega}$ and the value of $B\Omega(X^H, Y^H)$ on P, a direct computation leads finally to the result.□

At this point, we can assume, without loss of generality, that the "model" structure (u, η, ξ) on $\mathbf{R}^n$ is just that given in the Example above, and then consider on $gl(n, \mathbf{R})$ the almost contact structure (u', η', ξ') defined in the same example. With these hypotheses, $N_{\tilde{u}}$ is zero and hence Theorem 8.5.11 provides on P a weaker identity:

$$B\tilde{\Theta}(X^H, Y^H) = \{(N_{\bar{\phi}} + 2d\bar{\eta} \otimes \bar{\xi})(X, Y)\}^H . \tag{8.17}$$

These observations lead us to rediscover two well known results from the theory of almost contact manifolds.

Corollary 8.5.12 *Let $(\bar{\phi}, \bar{\eta}, \bar{\xi})$ be an almost contact structure on M. If there is on M an adapted torsion free connection, then the almost contact structure is normal.*

Proof. Without loss of generality, we assume $(\bar{\phi}, \bar{\eta}, \bar{\xi})$ "modeled" by (u, η, ξ) as given in the Example above, and consider (u', η', ξ') on $gl(n, \mathbf{R})$ as given in the same example. Then, we construct on FM the almost complex structure $\tilde{J}$ associated to them and to the torsion free connection. Then (8.17) holds and, since $\Theta = 0$ implies $\tilde{\Theta} = 0$, the result follows because the almost contact structure is normal if and only if $N_{\bar{\phi}} + 2d\bar{\eta} \otimes \bar{\xi}$ vanishes identically.□

Corollary 8.5.13 *If ω is a connection adapted to the almost contact structure $(\bar{\phi}, \bar{\eta}, \bar{\xi})$ on M, and if T is the torsion tensor of ω, then*

$$\begin{aligned}(N_{\bar{\phi}} + 2d\bar{\eta} \otimes \bar{\xi})(X, Y) &= -T(\bar{\phi}X, \bar{\phi}Y) + \bar{\phi}T(\bar{\phi}X, Y) \\ &\quad +\bar{\phi}T(X, \bar{\phi}Y) + T(X, Y) .\end{aligned}$$

Proof. Recalling the arguments in the proof of Corollary 8.5.12, a direct computation proves the identity.□

Let G, G', be positive definite inner products on $\mathbf{R}^n$ and $gl(n, \mathbf{R})$, respectively, and assume that they are compatible with the almost contact structures (u, η, ξ) and (u', η', ξ'), that is

(1) $G(u(\alpha_1), u(\alpha_2)) = G(\alpha_1, \alpha_2) - \eta(\alpha_1)\eta(\alpha_2)$;

(2) $G'(u'(A_1), u'(A_2)) = G'(A_1, A_2) - \eta'(A_1)\eta'(A_2)$,

for $\alpha_i \in \mathbf{R}^n$ and $A_i \in gl(n, \mathbf{R})$.

Then for any $\alpha \in \mathbf{R}^n$ and $A \in gl(n, \mathbf{R})$,

$$G(\alpha, \xi) = \eta(\alpha) \quad , \; G(\xi, \xi) = 1 \; ,$$

$$G'(A, \xi') = \eta'(A) \quad , \; G'(\xi', \xi') = 1 \; .$$

If we now consider on FM the Riemannian metric $\tilde{G}$ associated to G and G' according to the construction in Section 8.4, then we can state for $(\tilde{J}, \tilde{G})$ on FM the following theorem.

Theorem 8.5.14 *$(\tilde{J}, \tilde{G})$ is an almost Hermitian structure on FM; that is*

$$\tilde{G}(\tilde{J}X, \tilde{J}Y) = \tilde{G}(X, Y)$$

for all vector fields X, Y on FM.

Proof. Direct from the definitions.□

Now it is natural to study the Kähler form of the almost Hermitian structure $(\tilde{J}, \tilde{G})$, given by $\tilde{F}(X, Y) = \tilde{G}(X, \tilde{J}Y)$.

To make such study possible, we shall introduce some prior conditions on our algebraic "model" $\mathbf{R}^n \times gl(n, \mathbf{R})$ endowed with its Lie algebra structure and the almost complex structure $\tilde{u}$ associated to (u, η, ξ) and (u', η', ξ').

Consider first on $\mathbf{R}^n \times gl(n, \mathbf{R})$, the inner product $\tilde{G} = G \times G'$. A direct computation proves the following lemma.

Lemma 8.5.15 *$\tilde{G}$ is Hermitian with respect to $\tilde{u}$, that is*

$$\tilde{G}(\tilde{u}(\alpha_1, A_1), \tilde{u}(\alpha_2, A_2)) = \tilde{G}((\alpha_1, A_1), (\alpha_2, A_2)) \; ,$$

$(\alpha_i, A_i) \in \mathbf{R}^n \times gl(n, \mathbf{R})$.

We can define the fundamental form of $(\tilde{u}, \tilde{G})$ as the 2–form

$$\tilde{F} \in \Lambda^2(\mathbf{R}^n \times gl(n, \mathbf{R}))^* \; ,$$

given by

$$\overline{F}((\alpha_1, A_1), (\alpha_2, A_2)) = \tilde{G}((\alpha_1, A_1), \tilde{u}(\alpha_2, A_2)) ,$$

and if we denote by d the Chevalley-Eilenberg differential in the Lie algebra $\mathbf{R}^n \times gl(n, \mathbf{R})$, then $d\overline{F} \in \Lambda^3(\mathbf{R}^n \times gl(n, \mathbf{R}))^*$ is given by

$$\begin{aligned} 3d\overline{F}((\alpha_1, A_1),\ (\alpha_2, A_2), (\alpha_3, A_3)) & \\ = -\overline{F}([(\alpha_1, A_1), (\alpha_2, A_2)], (\alpha_3, A_3)) & \\ -\overline{F}([(\alpha_2, A_2), (\alpha_3, A_3)], (\alpha_1, A_1)) & \\ -\overline{F}([\alpha_3, A_3), (\alpha_1, A_1)], (\alpha_2, A_2)) . & \end{aligned}$$

Now we introduce a new definition (compare with Definition 8.4.4).

Definition 8.5.16 *If*

$$\alpha \in \Lambda^2(FM, \mathbf{R}^n \times gl(n, \mathbf{R})) ,$$

then

$$\tilde{\alpha} \in \Lambda^3(FM, \mathbf{R})$$

is the 3-form on FM given by

$$3\tilde{\alpha}(X, Y, Z) = \sum_{X,Y,Z} \tilde{G}(\alpha(X, Y), \tilde{u}(\theta Z, \omega Z))$$

for all vector fields X, Y, Z on FM ($\sum_{X,Y,Z}$ means cyclic sum).

Now, considering again $d\theta$ and $d\omega$ as $\mathbf{R}^n \times gl(n, \mathbf{R})$–valued forms, then $\widetilde{d\theta}, \widetilde{d\omega} \in \Lambda^3(FM, \mathbf{R})$, and we obtain

Theorem 8.5.17 *Let $\tilde{F}$ be the Kähler form of the almost Hermitian structure $(\tilde{J}, \tilde{G})$ on FM. Then*

$$d\tilde{F} = 2\widetilde{d\theta} + 2\widetilde{d\omega} .$$

Proof. Straightforward computation, taking into account that, for any of the basic cases, any term of the form $X\tilde{F}(Y, Z)$ is zero because $\tilde{F}(Y, Z)$ is a constant function on FM.□

Essentially the identity in Theorem 8.5.17 is similar to that in Theorem 8.4.5 but here we can make explicit in $d\tilde{F}$ the curvature and the torsion forms of ω. This will be possible through the Nijenhuis torsion $N_{\tilde{J}}$ and, again, the influence of the algebraic properties of the "model" on the geometry of FM is clear.

Lemma 8.5.18 *Let $\overline{F}$ be the fundamental form of $(\tilde{u}, \bar{G})$. Then*

$$2(\widetilde{\omega \wedge \theta} + \widetilde{\omega \wedge \omega}) = d\overline{F} \cdot (\theta, \omega) \ ,$$

where $d\overline{F} \cdot (\theta, \omega)$ is the differential of the 3-form on FM given by

$$(d\overline{F} \cdot (\theta, \omega))(X, Y, Z) = d\overline{F}((\theta X, \omega X), (\theta Y, \omega Y), (\theta Z, \omega Z)) \ .$$

Proof. Direct computation.□

Combining Theorem 8.5.17 with Lemma 8.5.18, and using the structure equations of ω, one obtains

Theorem 8.5.19 $d\tilde{F} = 2\tilde{\Theta} + 2\tilde{\Omega} + d\overline{F} \cdot (\theta, \omega)$.

Obviously, $d\overline{F} \cdot (\theta, \omega)$ vanishes if and only if $d\overline{F}$ does so. Hence,

Corollary 8.5.20 *If the fundamental form $\overline{F}$ of $(\tilde{u}, \tilde{G})$ is closed, then*

$$d\tilde{F} = 2\tilde{\Theta} + 2\tilde{\Omega} \ .$$

8.6 Application to spacetime structure

In Riemannian geometry the Hopf–Rinow theorem tells us that a manifold is geodesically complete if and only if it is Cauchy complete as a metric space in the infimum metric for arc length. For pseudo–Riemannian manifolds, and in particular for spacetime, the absence of a positive definite distance function makes the handling of incompleteness more difficult.

Indeed, there are spacetimes which are geodesically complete but which contain inextensible, incomplete timelike curves of bounded acceleration. Particles or observes following such trajectories would disappear from the universe, for example in a black hole, or suddenly appear for example from a white hole.

For further details see Hawking and Ellis [42]. Schmidt [80] suggested a way to handle spacetime incompleteness, by lifting the problem to the frame bundle and there using a Levi–Civita connection–related metric.

Definition 8.6.1 *The connection metric g_Γ on the frame bundle of a manifold with connection Γ is*

$$g_\Gamma = \theta \bullet \theta + \hat{\omega}_\Gamma \bullet \hat{\omega}_\Gamma : TFM \times TFM \longrightarrow \mathbf{R}$$

where θ is the canonical 1-form on FM, $\hat{\omega}_\Gamma$ is the $\mathbf{R}^{n^2}$-valued connection 1-form of Γ and $\bullet$ denotes the standard inner product on $\mathbf{R}^n$ and $\mathbf{R}^{n^2}$. Normally we are interested in g_Γ on a connected component of FM but we shall omit reference to this.

Example

Take $M = S^1$ with constant connection Γ having $\Gamma^1_{11} = \lambda \neq 0$. Then the connection metric g_Γ at $(x, b) \in FS^1$ is given by

$$g_\Gamma((p,q),(r,s)) = pr/b^2 + (q + \lambda bp)(s + \lambda br)/b^2 \ .$$

In this space (FS^1, g_Γ), the curve

$$\mathbf{c}: t \mapsto (-t \operatorname{mod} 1, e^{\lambda t}) \quad \text{for } t \in [0, t_m]$$

has the norm of its tangent vector equal to $\lambda e^{-\lambda t}$.

Hence the length of $\mathbf{c}$ is $(1 - e^{-\lambda t_m})/|\lambda|$ which remains finite even as $t_m \to \infty$. We observe that such a curve is the horizontal lift of one which circulates around S^1 t_m times. It follows that (FS^1, g_Γ) is incomplete (unless $\lambda = 0$). In fact, this space is essentially uniformly equivalent to

$$\{(u,v) \in \mathbf{R}^2 \mid |v| > 1/|\lambda| \}$$

in the standard metric. (Cf. Dodson and Sulley [26].)

Theorem 8.6.2 *(Properties of the connection metric on FM)*
(i): *g_Γ is unique up to uniform equivalence.*
(ii): *The action of $Gl(n, \mathbf{R})$ on (FM, g_Γ) is uniformly continuous.*

Proof. (i) Follows from the uniform equivalence of norms on a finite dimensional vector space.

(ii) Follows from the definition of g_Γ and the observation that if $f \in Gl(n, \mathbf{R})$ then for some $r > 0$ we have

$$\|f(X)\| \leq r\|X\| \quad \text{, for } X \in \mathbf{R}^n \ .$$

For in matrix components, if $u \in FM$, $Y \in T_u FM$, $f \in Gl(n, \mathbf{R})$ then

$$\begin{aligned} \theta(Y) &= b^{-1}X \ , \\ \theta(DR_f(Y)) &= (fb)^{-1}X = b^{-1}f^{-1}X \ , \\ \hat{\omega}_\Gamma(Y) &= (B + b\Gamma X)b^{-1} \ , \\ \hat{\omega}_\Gamma(DR_f(Y)) &= f(B + f\Gamma X)b^{-1}f^{-1} \ . \end{aligned}$$

Then

$$\|DR_f(Y)\| \leq m\|Y\| \quad \text{for some } m > 0 \ . \square$$

Corollary 8.6.3 *Every action R_f of $f \in Gl(n, \mathbf{R})$ admits, by uniform continuity, a unique uniformly continuous extension $\bar{R}_f$ to the Cauchy completion $(\overline{FM}, \bar{g}_\Gamma)$ in which (FM, g_Γ) is dense.*

Corollary 8.6.4 *Fibres of FM are homogeneous spaces and hence complete Riemannian submanifolds of (FM, g_Γ).*

This construction gives us a new criterion for completeness of any manifold with connection:

Definition 8.6.5 *A manifold M with a linear connection Γ is called* b*-incomplete or connection incomplete if (FM, g_Γ) is complete.*

Then a curve **c** in M with connection Γ is b–incomplete if its horizontal lift to FM is finite in length and inextensible in (FM, g_Γ).

In the spacetime situation, this is intuitively attractive because the length of the horizontal lift in FM corresponds to using in M, a Euclidean norm on the tangent vector to **c**, with respect to a parallelly propagated pseudo–orthonormal frame.

The next result shows that we get the right answer when we do have the Hopf–Rinow Theorem.

Theorem 8.6.6 *(Riemannian completeness) If (M, g) is a Riemannian manifold then (M, g) is complete if and only if (FM, g_Γ) is complete, where Γ is the Levi–Civita connection of g.*

Proof. (Cf. Schmidt [80] and [22]) By definition, $M = FM/G$ and it is straightforward to show that G acts uniformly continuously on (FM, g_Γ). If (FM, g_Γ) is complete then in particular so are integral curves of its horizontal vector fields, and these correspond to geodesics in (M, g); so (M, g) is complete.

Conversely, if (FM, g_Γ) is incomplete, then we can show that each Cauchy limit point is accessible on a horizontal curve. But this curve must be incomplete in FM and therefore projects onto an incomplete geodesic in M.□

A detailed study of the geometry of (FM, g_Γ) for spacetime manifolds was given by Dodson [22]. The interest stems from the fact that a b–incomplete spacetime can be completed to give a topological space in which the original spacetime is dense. By this device it is possible to give meaning to the notion of being "near to a singularity", which is important in the study of physical fields in their vicinity. Also in [22] are details of the b–completion of the 2–dimensional Friedmann spacetime, carrying the essential features of most widely accepted evolutionary models of the universe.

Theorem 8.6.7 *(Spacetime Completion) Consider a spacetime manifold (M,g), so g is a Lorentz type metric, such that M is* b*–incomplete with respect to the Levi–Civita connection. Then M admits a completion $\overline{M}$ as a topological space in which M is dense. The construction holds for any pseudo–Riemannian manifold.*

Proof. Take a connected component OM of the pseudo–orthonormal bundle of frames, with structure group the Lorentz group L.

Then L acts uniformly continuously on (OM, g_Γ) and the action extends uniformly continuously to the completion $\overline{OM}$. Define $\overline{M} = \overline{OM}/L$, and it follows that $M = OM/L$ is dense in $\overline{M}$.□

Corollary 8.6.8 *We observe that $\overline{M}$ is essentially unique because g_Γ is unique up to uniform equivalence.*

We call $\partial M = \overline{M} \setminus M$ the **b**–*boundary* of (M, g) and $\overline{M}$ the **b**–*completion* of M; $\overline{M}$ need not be Hausdorff, by the following.

Example. The **b**–completion of S^1, with constant $\Gamma^1_{11} = \lambda \neq 0$ connection, consists of S^1 and an outside point whose only neighbourhood is the whole of $\overline{S^1}$.

In fact $\overline{M}$ need not be locally compact either:

Example. Let M be the connected component of $(0, -2)$ in

$$\{(x, \sin 1/x) \in \mathbb{R}^2 \mid x \neq 0\} \cup \{(0, y) \mid |y| \leq 1\} .$$

Then $\partial M = \{(x, \sin 1/x) \in \mathbb{R}^2 \mid x \neq 0\} \cup \{(0, -1)\}$ and $(0, -1)$ has no compact neighbourhood.

Note that a sufficient condition for **b**–incompleteness of M is geodesic incompleteness, but not conversely.

Despite the lack of Hausdorffness and of local compactness that may occur for the **b**–completion of a spacetime, the notion of **b**–incompleteness still remains the favoured criterion for the existence of a spacetime singularity. For in any reasonable spacetime it turns out that on a curve which ends on the **b**–boundary it is very likely that curvature, and hence tidal forces, will become unbounded. We summarize a few properties in the following.

Theorem 8.6.9 *Let (M, g) be a pseudo–Riemannian manifold with Levi–Civita* b*–completion $\overline{M} = \overline{OM}/G$ and denote by $\overline{p}$ the projection of $\overline{OM}$ onto $\overline{M}$. Then we have:*

(i): *the action of G is transitive on the fibres of $\overline{p}$ and $\overline{p}$ is an open map;*

(ii): *the fibres of $\overline{p}$ are complete in the induced metric;*

(iii): *$\overline{M}$ is not T_1 if the orbits of G are not closed in $\overline{OM}$;*

(iv): *$\overline{M}$ is T_2 if and only if the graph of G is closed in $\overline{OM} \times \overline{OM}$;*

(v): *$\partial M = \overline{M} \setminus M$ contains the topological boundary;*

(vi): *every point in ∂M is the projection by $\overline{p}$ of the endpoint of a horizontal curve in OM.*

Proof. (Cf. Dodson [22])

(i) This is straightforward and forces homogeneity on the fibres, and hence **(ii)** follows.

(iii) This is straightforward.

(iv) Observe that the graph of G is the equivalence relation

$$\Delta = \{(u, \overline{R}_f(u)) \,|\, u \in \overline{OM} \ , \ f \in G\}$$

and $\overline{M} = \overline{OM}/\Delta$. Moreover, the identity on $\overline{M}$ has the graph $D = \{(x,x) \,|\, x \in \overline{M}\}$ with $\overline{M}/D = \overline{M}$ and

$$\Delta = (\overline{p} \times \overline{p})^{\leftarrow} D \ .$$

If $\overline{M}$ is Hausdorff and $(a,b) \notin \Delta$ then we can separate $\overline{p}(a)$ and $\overline{p}(b)$ by disjoint open U, V. Now, no point of $\overline{p}^{-1}U$ lies on a fibre that meets $\overline{p}^{-1}V$; so they are not Δ-related. Therefore, $\overline{p}^{-1}U \times \overline{p}^{-1}V$ is an open neighbourhood of (a,b), disjoint from Δ. So Δ is closed. The converse follows by openness of $\overline{p}$.

(v) Take U an open submanifold of M with its closure $\tilde{U}$ compact in M; then for all $x \in \overset{\circ}{U} = \tilde{U} \setminus (\textit{interior}\ U)$ we know $x \notin U$ since U is open and we can construct a curve $\mathbf{c}$ in U with endpoint x. Through any $b_0 \in \overline{p}^{-1}(\mathbf{c}(0))$ construct the unique horizontal lift $\overline{\mathbf{c}}$ of $\mathbf{c}$. Then $\overline{\mathbf{c}}$ has endpoint $b_1 \in \overline{OM}$, by compactness. So, for all large enough $n \in \mathbb{N}$, $\overline{\mathbf{c}}$ eventually remains in the open ball centre b_1 radius $1/n$, in $\overline{OM}$. Hence b_1 can be connected to a point on $\overline{\mathbf{c}}$ in OM by a minimizing geodesic of finite length. This geodesic projects onto a **b**-incomplete curve in U with endpoint $x \in \partial U$. Therefore $\overset{\circ}{U} \subseteq \partial U$.

(vi) Every point in ∂M is represented by a class of Cauchy sequences in (OM, g_Γ), it can be shown that an equivalent sequence exists on a horizontal curve.□

Some recent work on spacetime singularities has been concerned with *stability* of incompleteness (cf. Canarutto and Dodson [8], Dodson [23], Del Riego and Dodson [21]). We discuss this in the next Chapter, cf. 9.5.

Chapter 9

Systems of connections and universality

Introduction

In geometry and its applications there are commonly encountered situations that require analysis not just of one connection but of a whole family of connections on a given space. This leads to difficulty because the space of all connections on any given space is not even locally finite–dimensional. The difficulty is overcome to a considerable extent by the use of systems of connections, for they allow the selection of pertinent finite–dimensional families of connections. Most importantly, on each system there is a natural connection which is universal with respect to the system; that is, every connection from the system is a pullback of the universal connection.

This approach has interesting consequences for geometry, topology, theoretical physics and statistical theory.

9.1 Connections on a fibred manifold

In the sequel we shall be interested mainly in linear connections but en route we shall need a more general definition to develop the notion of universal connection.

Definition 9.1.1 *A fibred manifold is a surjective submersion* $\mathsf{p}: E \longrightarrow B$ *and as such* p *has maximal rank everywhere. Sometimes such a* p *is called* a surmersion.

Every fibre bundle is a fibred manifold but not conversely. For example, the natural projection

$$\mathsf{p}: \mathbf{R}^2 - \{0\} \longrightarrow \mathbf{R} \ : (x, y) \longmapsto x$$

is a fibred manifold but not a fibre bundle.

A connection on a fibred manifold

$$\mathbf{p}: E \longrightarrow B$$

is a section Γ of the first jet bundle $\mathbf{q}: JE \longrightarrow E$.

Since JE consists of classes of sections of $\mathbf{p}$ that are equivalent up to first derivative, a typical element of JE at $X \in E$ is represented by a linear map, the derivative

$$T\sigma: T_{\mathbf{p}(X)}B \longrightarrow T_X E$$

of a section σ.

Moreover, since σ is a section then $T\mathbf{p} \circ T\sigma = 1_{TB}$ and hence our representative $T\sigma$ corestricts to the identity on the subbundle $TB \hookrightarrow TE$.

Proposition 9.1.2 *$JE \longrightarrow E$ is an affine subbundle of the vector bundle*

$$T^*B \otimes_B TE\ .$$

Proof. It is precisely that subbundle which projects onto the identity 1_{TB} when viewed as a section of TE.□

Local expressions

The local coordinate expression of a connection Γ on a fibred manifold $\mathbf{p}: E \to B$ is easily seen to be of the form

$$\Gamma: E \longrightarrow JE \hookrightarrow T^*B \otimes_B TE$$
$$(x^i, X^\alpha) \longmapsto dx^i \otimes \frac{\partial}{\partial x^i} - \Gamma_i^\alpha\, dx^i \otimes \frac{\partial}{\partial X^\alpha}$$

where

$$\Gamma_i^\alpha = \frac{\partial}{\partial x^i}\gamma^\alpha$$

and (γ^α) represents the class of sections of E determined by Γ. The minus sign is introduced as a convenient convention.

Each connection induces a splitting into horizontal and vertical distributions:

$$TE \longrightarrow HE \oplus_E VE$$
$$(x^i, X^\alpha, \dot{x}^i, \dot{X}^\alpha) \to (x^i, X^\alpha, \dot{x}^i, \dot{x}^i\Gamma_i^\alpha) \oplus (x^i, X^\alpha, 0, \dot{X}^\alpha - \dot{x}^i\Gamma_i^\alpha)$$

yielding horizontal and vertical vector subbundles.

On a principal G–bundle the horizontal distribution is required to be G–invariant.

Examples of linear connections

When $E \longrightarrow B$ is the tangent bundle $TM \longrightarrow M$ then our connection appears in the form

$$\Gamma: TM \longrightarrow JTM \hookrightarrow T^*M \otimes_{TM} TTM$$

$$(x^i, \dot{x}^i) \longmapsto dx^i \otimes \frac{\partial}{\partial x^i} - \dot{x}^j \Gamma^k_{ji}\, dx^i \otimes \frac{\partial}{\partial \dot{x}^k}$$

When $\mathbf{p}: E \longrightarrow B$ coincides with the frame bundle $FM \longrightarrow M$ then our connection appears in the form

$$\Gamma: FM \longrightarrow JFM \hookrightarrow T^*M \otimes_{FM} TFM$$

$$(x^i, X^k_j) \longmapsto dx^i \otimes \frac{\partial}{\partial x^i} - \Gamma^k_{ji} dx^i \otimes \frac{\partial}{\partial X^k_j}$$

Definition 9.1.3 *The connection* 1*-form of a connection* Γ *on a fibred manifold* $\mathbf{p}: E \longrightarrow B$ *is the vertical vector* 1*-form*

$$\omega_\Gamma: E \longrightarrow T^*E \otimes_E VE$$

$$(x^i, X^\alpha) \longmapsto (d\dot{X}^\alpha - \Gamma^\alpha_i\, d\dot{x}^i) \otimes \frac{\partial}{\partial \dot{X}^\alpha}\ .$$

In the particular case of the frame bundle $FM \longrightarrow M$, the vertical fibres of TFM are isomorphic to the Lie algebra of the structure group $G = Gl(n, \mathbf{R})$. Hence we can express the vertical 1–form ω_Γ as an $\mathbf{R}^{n^2}$–valued form and furthermore obtain a conveniently scaled connection form

$$\hat{\omega}_\Gamma: TFM \longrightarrow \mathbf{R}^{n^2}$$

$$(x^i, X^k_j, \dot{x}^i, B^k_j) \longmapsto (B^k_j - X^k_j \Gamma^k_{im} \dot{x}^m)(X^k_j)^{-1}$$

Notation for sections

For any fibred manifold $\mathbf{p}: E \longrightarrow B$ we shall denote by $Sec\,(E/B)$ the set of sections of the surjection $\mathbf{p}$. Accordingly, $Sec\,(JE/E)$ is the set of connections on the given fibred manifold. On a G–bundle $E \longrightarrow E/G$ we shall denote the set of connections by $Con\,(E/G)$. One of the main results we describe is the bijection

$$Sec\,(JE/E) \longleftrightarrow Con\,(E/G)\ .$$

9.2 Principal bundle connections

The motivation for the ideas in this Chapter comes from the work of P.L. García [37] and we formulate three of his theorems as follows.

Theorem 9.2.1 *(Principal bundle lifting) In the diagram*

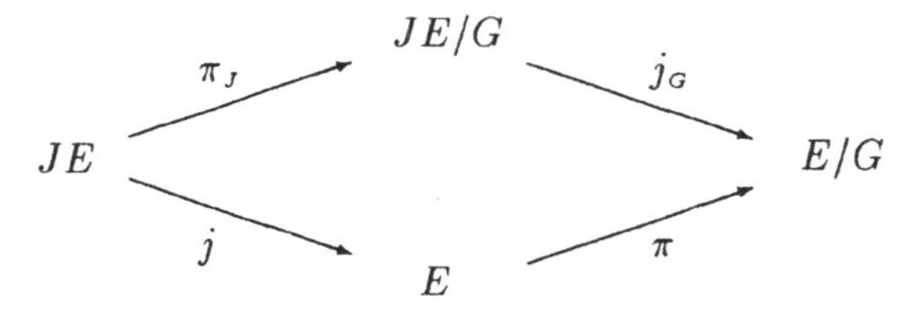

let the bottom line be the first jet bundle over a principal $G-$ bundle. Then the group action lifts to JE, $\pi \circ j$ is a fibred bundle, π_J is a principal G–bundle, and up to isomorphism JE is the pullback $j_G^* E$.

Proof. **(i)** The action lifts to a free action on the right as follows:

$$JE \times G \longrightarrow JE \ : (\bar{s}_x, a) \mapsto (\overline{a \cdot s})_x = \bar{s} \cdot a \ .$$

Now, if $\bar{s}_x = \bar{s}'_x$ then

$$(a \cdot s)(x) = s(x) \cdot a = s'(x) \cdot a = (a \cdot s')(x) \ ,$$

$$T(a \cdot s)_x = Ta_{s(x)} \cdot Ts_x = Ta_{s'(x)} \cdot Ts'_x = T(a \cdot s')_x \ .$$

Also,

$$\bar{s}_x \cdot (a \cdot b) = (\overline{(a \cdot b) \cdot s})_x = (\overline{a \cdot s})_x \cdot b = (\bar{s}_x \cdot a) \cdot b \ ,$$

$$\bar{s}_x \cdot a = \bar{s}_x \ \Rightarrow \ (\overline{a \cdot s})_x = \bar{s}_x \ \Rightarrow \ s(x) \cdot a = s(x) \ \Rightarrow \ a = 1 \ .$$

The action can be shown to be smooth.

(ii) Let $x_0 \in E/G$, $\sigma: U \longrightarrow E$ a local section of π for some open neighbourhood U of x_0. Then we get a commutative diagram

$$\begin{array}{ccc} j^{\leftarrow}\sigma U & \longrightarrow & j_G^{\leftarrow} U \\ {\scriptstyle j}\big\downarrow & & \big\downarrow{\scriptstyle j_G} \\ \sigma U & \stackrel{\sigma}{\longleftarrow} & U \end{array} \qquad \text{with } j_\sigma(\bar{s}_x) = \bar{s}_x \cdot G \ .$$

Since σ is a section it is a local diffeomorphism and induces by j_σ a fibre bundle structure on $j_G^{\leftarrow} U$. This structure turns out to be independent of the choice of σ, and differentiable.

(iii) We need to show that π_J is differentiable and gives a fibring of JE, then it is a principal G–bundle fibring. Take $x_0 \in E/G$ and local section σ as before.

Then the composite

$$\pi^{\leftarrow}U \xrightarrow{\pi} U \xrightarrow{\sigma} E$$

determines a unique differentiable map

$$\mathbf{g}: \pi^{\leftarrow}U \longrightarrow G \ : s \mapsto \mathbf{g}(s) \ \text{ with } \ s \cdot \mathbf{g}(s) = \sigma \cdot \pi(s) \ .$$

Now consider the following diagram, which commutes:

$$\begin{array}{ccccc}
j^{\leftarrow}(\pi^{\leftarrow}U) & \xrightarrow{i} & j^{\leftarrow}(\pi^{\leftarrow}U) \times \pi^{\leftarrow}U & \xrightarrow{g_*} & JE \times G \\
& \searrow^{\pi_J} & & & \downarrow \pi_1 \\
& & j_G^{\leftarrow}U & \xrightarrow{j_\sigma^{-1}} & JE
\end{array}$$

with $\mathbf{g}_*(\bar{s}_x, t) = (\bar{s}_x, \mathbf{g}(t))$. This traps π_J into being locally differentiable, and hence globally so since x_0 was arbitrary.

The local triviality is satisfied by putting

$$\varphi: j^{\leftarrow}(\pi^{\leftarrow}U) \longrightarrow j_G^{\leftarrow}U \times G \ : \bar{s}_x \mapsto (\pi_J(\bar{s}_x), \mathbf{g}(s)^{-1}) \ .$$

(iv) The pullback diagram is

$$\begin{array}{ccc}
j_G^* E & \longrightarrow & E \\
\downarrow & & \downarrow \pi \\
JE/G & \xrightarrow{j_G} & E/G
\end{array}$$

and evidently the correspondence of JE with $j_G^* E$ is given by

$$\bar{s}_x \longmapsto (\bar{s}_x \cdot G, s(x)) \ . \square$$

This theorem shows that the first jet functor J, which is defined on the category of fibred manifolds, actually restricts to be a functor on the category of principal G–bundles; explicitly, we have $J\pi = \pi_J$ on the bundle maps.

The next result shows that our definition of connections on a fibred manifold gives the right answer on principal bundles.

Theorem 9.2.2 *(Characterising principal connections)* *Let*

$$JE \xrightarrow{j} E \xrightarrow{\pi} E/G = M$$

be the first jet bundle over a principal G–bundle. Then the set of connections on the principal bundle is in bijective correspondence with sections of the jet bundle:

$$Sec\,(JE/E) \longleftrightarrow Con\,(E/G)\ .$$

Proof. Take any $\tilde{\gamma} \in Sec\,(JE/E)$ and define the distribution

$$\Gamma_\gamma : e \longmapsto T\gamma(T_x M)$$

where $\tilde{\gamma}(e) = \bar{\gamma}_x$, $\gamma(x) = e$, for each $e \in E$ and $\pi(e) = x$.

Now, since each representative γ of $\bar{\gamma}_x$ is a local section, the distribution has the correct dimension and is properly transverse to the fibres of j.

Let $e \in E$ and $a \in G$. By construction,

$$\tilde{\gamma}(e \cdot a) = \tilde{\gamma}(e) \cdot a$$

so

$$Ta\Gamma_\gamma(e) = \Gamma_\gamma(e \cdot a)$$

and the distribution is G–invariant, and therefore a (principal) connection on the principal G–bundle. It is easy to see that any two members of $Sec\,(JE/E)$ that yield the same distribution are themselves the same. Finally, if Γ is the distribution of any connection on the principal G–bundle, then it determines the unique section

$$\tilde{\Gamma} : E \longrightarrow JE\ : e \mapsto \bar{s}_x \ \text{ with } \ \pi(e) = x$$

where $Ts(T_x M) = \Gamma$.□

We can take this characterization of connections a little further in a way that is convenient later, by pushing the correspondence back to G–invariant sections defined on the base. This is another exploitation of the nice functorial behaviour of J.

Theorem 9.2.3 *(Principal connections as sections over the base)* *Let*

$$JE \stackrel{j}{\longrightarrow} E \stackrel{\pi}{\longrightarrow} E/G$$

be the first jet bundle over a principal G–bundle. Then there are bijective correspondences

$$Con\,(E/G) \longleftrightarrow Sec\,(JE/E) \longleftrightarrow Sec\,((JE/G)/(E/G))\ .$$

Proof. The first of these bijections was established in the previous theorem. The second correspondence arises from the fact that JE is an isomorph of the pullback $j_G^* E$ by Theorem 9.2.1.□

Consider the diagram

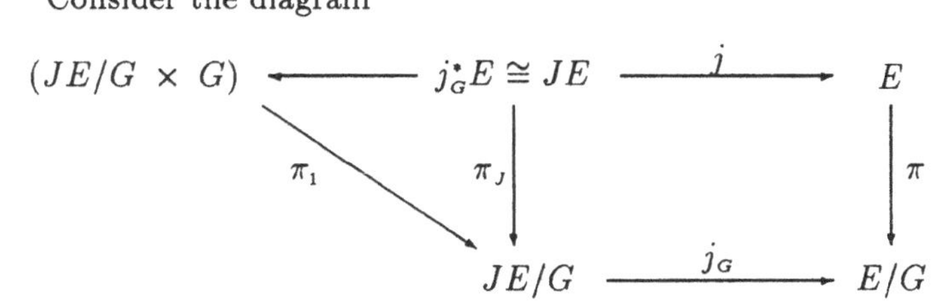

Evidently, any section $\tilde{\Gamma}$ of the surjection j_G induces a unique section Γ of the surjection j, simply by feeding $(\tilde{\Gamma}, 1_E)$ into the isomorphism with $j_G^* E$. Conversely, given $\Gamma \in Sec\,(JE/E)$ we have its correspondent $\Gamma^* \in Sec\,(j_G^* E/E)$ and so can put

$$\tilde{\Gamma}: E/G \longrightarrow JE/G \ : \bar{e} \mapsto \pi_1 \circ \Gamma^*(e) \ .$$

This is well-defined because of the fibrings, and it is a section because

$$j_G \circ \tilde{\Gamma}(\bar{e}) = j_G \circ \pi_1 \circ \Gamma^*(e) = j_G \circ \pi_J \circ \Gamma(e) = \pi \circ j \circ \Gamma(e) = \bar{e} \ .$$

Summary for the principal bundle of frames

The three results in this section have immediate application to the frame bundle $FM \longrightarrow M$. In summary we have the following:

$$\begin{array}{ccccc}
(JFM/G \times FM) & \longleftarrow & j_G^* FM \cong JFM & \stackrel{j}{\longrightarrow} & FM \\
& \searrow^{\pi_1} & \downarrow{\scriptstyle \pi_J} & & \downarrow{\scriptstyle \pi} \\
& & JFM/G & \stackrel{j_G}{\longrightarrow} & M \cong FM/G
\end{array}$$

(i):

$$Con\,(M) \longleftrightarrow Sec\,(JFM/FM) \longleftrightarrow Sec\,((JFM/G)/M) \ ;$$

(ii): J is principal–G–bundle–preserving, $J\pi = \pi_J$.

It is an easy exercise to establish the following, as pointed out by Canarutto and Dodson [8].

Proposition 9.2.4 *For the frame bundle $\pi_F: FM \longrightarrow M$*

(i): *$JFM/G \longrightarrow M$ is the affine subbundle*

$$((1_{T^*M} \otimes T\pi_F)^{\leftarrow} 1_{TM})/G \hookrightarrow T^*M \otimes TFM/G$$

whose associated vector bundle is $T^*M \otimes_M VFM/G$, where VFM is the vertical subbundle of TFM.

(ii): *$VFM/G \cong M \times \mathfrak{g}$ where $\mathfrak{g}$ is the Lie algebra of G, and there is one isomorphism induced for each choice of point in FM.*

9.3 Systems of connections

Observe that in the proof of Theorem 9.2.3 the affine bundle JE was parametrized by part of the product bundle $(JE/G \times E)$, namely by $j_G^* E$. Moreover, the full product is actually the fibred product over the base E/G, and it is a trivial G-bundle over JE/G. We have seen that each section of JE/G over E/G determines a unique connection, and conversely. Hence it is natural to introduce the following

Definition 9.3.1 *The system of all connections on a principal G-bundle*

$$E \longrightarrow E/G$$

consists of the fibred morphism over E:

$$\xi: JE/G \times_{E/G} E \longrightarrow JE \ : (\bar{s}_x \cdot G, e) \mapsto \bar{s}_x$$

where we view JE as a subbundle of $T^*(E/G) \otimes TE$ and $\bar{s}_x$ appears as the linear map

$$(Ts)_x: T_x(E/G) \longrightarrow T_{s(x)}E \ .$$

More generally, a *system of connections* on a fibred manifold

$$E \longrightarrow B$$

consists of a fibred morphism over E

$$\eta: C \times_B E \longrightarrow JE \hookrightarrow T^*B \otimes_B TE$$

where $C \longrightarrow B$ is a fibred manifold and we call C *the system space*. Then any section of $C \longrightarrow B$ determines a unique connection on $E \longrightarrow B$ as a section of $JE \longrightarrow E$. This idea was introduced by Mangiarotti and Modugno [61] and originally it was called a structure of connections. A comprehensive treatment has been given by Modugno [63].

Examples of systems of linear connections

Consider two of the many ways to see linear connections on a manifold M:

(i): as the system of all (linear) connections on the fibred manifold

$$\pi_T: TM \longrightarrow M$$

with system space

$$C_T = (1_{T^*M} \otimes_M T\pi_T)^{\leftarrow} 1_{TM} \subset T^*M \otimes_M JTM$$

where we view 1_{TM} as a section of $T^*M \otimes_M TM$ in $T^*M \otimes_{TM} TTM$.

(ii): as the system of all (principal) connections on the principal G–bundle

$$\pi_F: FM \longrightarrow M$$

with system space

$$C_F = JFM/G \hookrightarrow T^*M \otimes_{TM} TFM/G\ .$$

In fact these two representations are equivalent because of the following (cf. Del Riego and Dodson [21]).

Theorem 9.3.2 *(Characterising linear connections) For any manifold M there are bijections between*

(a) $Con(FM/G) = Con(M) = Sec(JFM/FM)$;

(b) $Sec((JFM/G)/M) = Sec(C_F/M)$;

(c) $Sec(JTM/TM)$;

(d) $Sec(C_T/M)$.

Proof. We have already proved **(a)** $\Leftrightarrow$ **(b)** in Theorem 9.2.3, and **(c)** $\Leftrightarrow$ **(d)** follows from 9.1.

The map

$$C: TFM/G \longrightarrow JTM \hookrightarrow T^* \otimes_{TM} TTM$$
$$(x^i, \dot{x}^i, B^i_j) \mapsto \big((x^i, X^i) \mapsto (x^i, \dot{x}^i, X^i, X^i B^k_j)\big)$$

induces a smooth bijection **(b)** $\Leftrightarrow$ **(d)** by means of the composition

$$JFM/G = C_F \hookrightarrow T^*M \otimes_{TM} TFM \xrightarrow{1\otimes\rho} C_T \hookrightarrow T^*M \otimes_M JTM\ .$$

Hence our bijection is

$$Sec(C_F/M) \longrightarrow Sec\ (C_T/M)\ : \tilde{\Gamma} \mapsto \bar{\Gamma} = 1 \otimes \rho \circ \tilde{\Gamma}\ .\square$$

Corollary 9.3.3 *Each connection Γ on $\pi_M: FM \longrightarrow M$ induces unique sections $\tilde{\Gamma} \in Sec\,(C_F/M)$ and $\bar{\Gamma} \in Sec\,(C_T/M)$, hence inducing diffeomorphisms*

$$FM \cong \tilde{\Gamma} \circ \pi_F(FM) \hookrightarrow C_F \times_M FM$$

$$TM \cong \bar{\Gamma} \circ \pi_T(TM) \subset C_T \times_M TM \ .$$

We observe that Theorem 9.3.2 recovers the classical result of Nomizu [72] which established the equivalence of linear connections on TM with principal connections on FM.

9.4 Universal Connections

Consider, on an arbitrary principal G–bundle $\mathbf{p}: E \longrightarrow E/G$ the vertical vector valued 1–form $\lambda \in T^*JE \otimes VE$ given locally by

$$\lambda: (X^i, Y^\alpha, E_i^\alpha) \longmapsto (0, Y^\alpha - E_i^\alpha X^i) \ .$$

García [37] pointed out that the kernel of λ is a G–invariant distribution and actually a (principal) connection. Explicitly, at each $\bar{s} \in JE$ with $j(\bar{s}) = s$,

$$\lambda_{\bar{s}} = (1_{TE} - Ts \circ T\mathbf{p}) \circ Tj \ ,$$

here $JE \xrightarrow{j} E \xrightarrow{\mathbf{p}} B$.

Now take any $\sigma \in Sec\ (JE/E)$ and consider

$$\sigma^*\lambda_{\bar{s}} = \lambda_{\bar{s}} \circ T\sigma = (1_{TE} - Ts \circ T\mathbf{p}) \circ Tj \circ T\sigma = 1_{TE} - Ts \circ T\mathbf{p} \ ,$$

at each $\bar{s}$ with $j(\bar{s}) = s$. This is evidently the connection 1–form of σ, here taking values in VE. Thus, σ is a pullback of λ and λ is in this sense universal so we have the following.

Theorem 9.4.1 *(Universal principal bundle connections) Let*

$$JE \xrightarrow{j} E \xrightarrow{\pi} E/G$$

be the first jet bundle over a principal G–bundle. Then on the principal G–bundle $\pi_J: JE \longrightarrow JE/G$ there is a (principal) connection Λ which has the universal property that for each (principal) connection Γ on $\pi: E \longrightarrow E/G$

$$\Gamma = \Gamma^*\Lambda \ .$$

Corollary 9.4.2 *If ω_Γ, Ω_Γ and ω_Λ, Ω_Λ are the respective connection 1-form and curvature 2-form for Γ and Λ then*

$$\omega_\Gamma = \Gamma^*\omega_\Lambda \quad \text{and} \quad \Omega_\Gamma = \Gamma^*\Omega_\Lambda \ .$$

Here $\Omega_\Gamma = (1/2)[\omega_\Gamma, \omega_\Gamma]$.

We can formulate the universal property in terms of sections of jet bundles as follows.

Theorem 9.4.3 *(Universal fibred manifold connections) Let*

$$JE \xrightarrow{j} E \xrightarrow{p} B$$

be the first jet bundle over a fibred manifold. Then there is a distinguished connection Λ on the fibred manifold

$$\pi_1 : JE \times E \longrightarrow JE \ ,$$

and having the universal property that

$$\Gamma = \Gamma^*\Lambda \ .$$

for any connection $\Gamma \in Sec\,(JE/E)$.

Proof. The construction is essentially the same as before. Observe that

$$JE \hookrightarrow T^*B \otimes TE$$

and similarly

$$J(JE) \hookrightarrow T^*E \otimes TJE \ ,$$

as affine subbundles. Put

$$\begin{aligned} \Lambda : JE \times_B E &\longrightarrow J(JE \times E) \hookrightarrow T^*(JE) \otimes T(JE \times E) \\ (\bar{s}_x, e) &\mapsto ((X, Y, S) \mapsto (X, Y, S, TsX)) \end{aligned}$$

Then $\Lambda \in Sec\,(J(JE \times E)/JE \times E)$ and is therefore a fibred manifold connection.

Take any $\Gamma \in Sec\,(JE/E)$, so we have a map

$$\begin{aligned} \Gamma : E &\longrightarrow JE \hookrightarrow T^*B \otimes TE \\ e &\longmapsto Ts^\Gamma \ . \end{aligned}$$

Then

$$T\pi_2 \circ \Lambda \circ (\Gamma, 1_E)(e)(X, Y, S) = \Gamma(e)X$$

that is,

$$\Gamma = \Gamma^*\Lambda . \square$$

Corollary 9.4.4 *$\omega_\Gamma = \Gamma^*\omega_\Lambda$ and $\Omega_\Gamma = \Gamma_\Lambda$.*

Next we can see how the universal connection fits into the framework of a system of linear connections.

Theorem 9.4.5 *(Universal frame bundle connection) For the frame bundle system of connections*

$$JFM/G \times_M FM \xrightarrow{\eta_F} JFM \hookrightarrow T^*M \otimes_{TM} TFM/G$$
$$([\bar{s}_x], b) \mapsto \left[T_xM \xrightarrow{[\bar{s}_x]} T_bFM\right]$$

each $\widetilde{\Gamma} \in Sec((JFM/G)/M)$ determines the unique principal connection $\Gamma \in Sec(JFM/FM)$ with

$$\Gamma = \eta_F \circ (\widetilde{\Gamma} \circ \pi_F \circ 1_{FM}) \; .$$

On the principal G-bundle

$$\pi_1 \colon JFM/G \times_M FM \longrightarrow JFM/G$$

there is a (principal) connection

$$\begin{aligned} \Lambda \colon JFM/G \times_M FM \; & \to J(JFM/G \times_M FM) \hookrightarrow \\ & \hookrightarrow T^*(JFM/G) \otimes_{TM} T(JFM/G \times_M FM) \\ (x^i, \gamma^k_{ij}, b^i_j) \; & \mapsto \left[(X^i, Y^k_{ij}) \mapsto (X^i, Y^k_{ij}, b^m_j \gamma^k_{mr} X^r)\right] \; . \end{aligned}$$

Moreover, Γ is the universal connection for the frame bundle system. Explicitly, each $\widetilde{\Gamma} \in Sec((JFM/G)/M)$ gives an injection $(\widetilde{\Gamma} \circ \pi_F, 1_{FM})$, of FM into $JFM/G \times FM$, which is a section of π_1 and Γ coincides with the restriction of Λ to this section:

$$\Lambda_{|(\widetilde{\Gamma} \circ \pi_F, 1_{FM})FM} = \Gamma \; .$$

Corollary 9.4.6 *The universal connection Λ for a frame bundle*

$$\pi_F \colon FM \longrightarrow M$$

induces a tangent vector field

$$\begin{aligned} \Sigma_\Lambda \colon JFM/G \times_M FM & \longrightarrow T(JFM/G \times_M FM) \\ (x^i, \gamma^k_{ij}, b^i_j) & \longmapsto (x^i, \gamma^k_{ij}, b^i_j, b^{ij}\gamma^k_{ij}, b^i_j \gamma^k_{im}, b^i_j) \end{aligned}$$

with $b^{ij} = b^i_j$.

There is an analogous formulation of the universal linear connection on the tangent bundle $\pi_T: TM \longrightarrow M$, and this is discussed in the context of sprays in Del Riego and Dodson [21].

Remark 9.4.7 Other universal connections were used in Narasimhan and Ramanan [70,71] (cf. [47,p.289] and [48,p.332]).

They proved that for any given Lie group G with a finite number of components and a given positive integer N, there is a principal G–bundle

$$E_G \longrightarrow E_G/G$$

carrying a connection Γ_G such that every connection Γ on any principal G–bundle $E \longrightarrow E/G$ with $dim\ E \leq N$ is the pullback of Γ_G to E.

A universal calculus exploiting the universal connection has been devised by Dodson and Modugno [25]; this appears to offer a convenient way to present gauge fields theories.

9.5 Applications

We give here four applications of the theory in this Chapter, one each in geometry, topology, physics and statistics.

Universal holonomy

Theorem 9.5.1 *(Universal holonomy group) Let $\pi_F: FM^0 \longrightarrow M$ be a connected component of a frame bundle over M with dim $M \geq 2$. Then all the holonomy bundles $\{JFM^0(\bar{s}_x) \mid \bar{s}_x \in JFM^0\}$ with respect to the universal connection Λ coincide with JFM^0. Hence, at each point $\bar{s}_x \in JFM^0$ the holonomy group of Λ is the structure group of FM^0.*

Proof. This is a special case of the result for principal bundles proved by García [37].

From Kobayashi and Nomizu [47,p.90] we can find a connection Γ on FM^0 such that all of its holonomy bundles coincide with FM^0. Take $\bar{s}_x, \bar{t}_y \in JFM^0$ and construct a Γ–horizontal curve

$$\mathbf{c}: [0,1] \longrightarrow FM^0 \text{ from } s(x) \text{ to } t(y) .$$

Also construct vertical curves in $j: JFM^0 \longrightarrow FM$:

$$\mathbf{d}: [0,1] \longrightarrow j^{\leftarrow}\{s(x)\} \quad \text{from } \bar{s}_x \text{ to } \Gamma \circ s(x) ,$$

$$\mathbf{h}: [0,1] \longrightarrow j^{\leftarrow}\{t(y)\} \quad \text{from } \Gamma \circ t(y) \text{ to } \bar{t}_y .$$

Then the composite path from $\bar{s}_x$ to $\bar{t}_y$ given by $\mathbf{d} \circ \Gamma \circ \mathbf{c} \circ \mathbf{h}$ is horizontal.□

Weil's Theorem

Theorem 9.5.2 *(Weil's theorem on characteristic classes) Take any frame bundle $\pi_F: FM \longrightarrow M$, or one of its subbundles, with structure group G. There is a well-defined algebra homomorphism into de Rham cohomology*

$$\mathbf{w}: I(G) \longrightarrow H^*(M, \mathbf{R})$$

where $I(G) = \sum_{k=0}^{\infty} I^k(G)$ and $I^k(G)$ is the real vector space of G-invariant k-linear maps $\mathfrak{g} \times \cdots \times \mathfrak{g} \longrightarrow \mathbf{R}$ defined on the Lie algebra of G.

Proof. (Cf. García [37], and Kobayashi and Nomizu [48,p.239].) The algebra structure on $I(G)$ is defined by

$$\begin{aligned} I^k(G) \times I^l(G) &\longrightarrow I^{k+l}(G) \\ (f, g) &\longmapsto fg \end{aligned}$$

where

$$(fg)(t_1, \ldots, t_{k+l}) = \frac{1}{(k+l)!} \sum_{\sigma} f(t_{\sigma(1)}, \ldots, t_{\sigma(k)}) g(t_{\sigma(k+1)}, \ldots, t_{\sigma(k+l)})$$

and σ is summed over all permutations of $(1, \ldots, k+l)$. Hence $I(G)$ is a real commutative algebra.

Let Λ be the universal connection for the system of all linear connections and let Ω_Λ be its curvature 2-form on JFM.

Take any $f \in I^k(G)$ and two linear connections Γ_1, Γ_2 on M with curvature 2-forms Ω_1, Ω_2 on FM respectively. Then from Theorems 9.4.3 and 9.4.5 we have

$$f(\Omega_1) = \Gamma_1^* f(\Omega_\Lambda) \; , \; f(\Omega_2) = \Gamma_2^* f(\Omega_\Lambda) \; .$$

But Γ_1 and Γ_2 lie in an affine subbundle of $T^*M \otimes_{TM} TFM$ and hence they are connected by a linear homotopy.

Now let Γ be any linear connection on M, with curvature 2-form Ω on FM. Define for each $f \in I^k(G)$ the $2k$-form on FM

$$f(\Omega)(V_1, \ldots, V_{2k}) = \frac{1}{(2k)!} \sum_{\sigma} sg(\sigma) f(\Omega(V_{\sigma(1)}, V_{\sigma(2)}), \ldots, \Omega(V_{\sigma(2k-1)}, V_{\sigma(2k)}))$$

where V_i are tangent to FM.

Next, G-invariance of f and the Bianchi identity allow $f(\Omega)$ to project onto a unique $2k$-form on M, which determines a class $\overline{f(\Omega)} \in H^{2k}(M, \mathbf{R})$. It is easily seen that the map $f \longmapsto \overline{f(\Omega)}$ is a real algebra homomorphism. It meets the properties required of $\mathbf{w}$ because any two linear connections on M are homotopic and so determine the same element in $H^{2k}(M, \mathbf{R})$.□

In fact this Theorem is true for any principal G-bundle and other proofs can be found in [48] based on Chern [10] and in [70] using universal connections.

Spacetime singularities

The next series of results lead to an application in spacetime singularity theory. We establish that the existence of an inextensible curve of finite proper length is stable under small conformal variations of the Lorentz structure. Physically, such a curve suggests a singularity since it indicates that a particle ceases to exist after a finite lifetime.

Our stability result is actually true for more general variations (see Theorem 9.5.6 below and Canarutto and Dodson [8]) but, physically, the conformal variations of the metric correspond to opening and closing of the null cone at each point. The result suggests that some general relativistic singularities are stable under variation of the background metric and hence are unlikely to be quantized away by any quantum theory of gravity. Previously it was shown by Gotay and Isenberg [39] that, under geometrical quantization, a massless Klein–Gordon scalar field on a positively curved spacetime could not escape the collapse of its state vector. Our method of establishing the result depends in an essential way on the universal linear connection.

Theorem 9.5.3 *(Connection metric on a frame bundle) Every linear connection Γ on a manifold M induces a Riemannian structure g_Γ on each connected component of the frame bundle. The universal connection Λ induces a symmetric bilinear form f_Λ on the system space $(JFM/G \times FM) = C_F$. The unique $\tilde{\Gamma} \in Sec\,(C_F/M)$ corresponding to a given linear connection Γ on M, induces an isometric embedding of (FM, g_Γ) in (C_F, f_Λ).*

Proof. Following Schmidt [80] and Marathe [62] we put

$$g_\Gamma = \theta \bullet \theta + \hat{\omega}_\Gamma \bullet \hat{\omega}_\Gamma : TFM \times TFM \longrightarrow \mathbf{R}$$

where θ is the canonical 1–form on FM, $\hat{\omega}_\Gamma$ is the $\mathbf{R}^{n^2}$–valued connection 1–form of Γ and $\bullet$ denotes the standard inner product on $\mathbf{R}^n$ and $\mathbf{R}^{n^2}$. (This is an example of an ω–associated tensor field of type $(0,2)$, cf. 8.4.)

Analogously, on C_F we construct

$$f_\Lambda = T\pi_2 \circ \theta \bullet T\pi_2 \circ \theta + \hat{\omega}_\Lambda \bullet \hat{\omega}_\Lambda : TC_F \times TC_F \longrightarrow \mathbf{R}\ .$$

Whereas g_Γ is evidently a Riemannian structure, f_Λ is degenerate on the fibres of $T\pi_2$.

However, each $\tilde{\Gamma} \in Sec\,(C_F/M)$ smoothly injects a copy of FM into C_F by the map $(\tilde{\Gamma} \circ \pi_F, 1_{FM})$ in Theorem 9.4.5. If this subspace is $S_\Gamma = (\tilde{\Gamma} \circ \pi_F, 1_{FM})$ then we find that the restriction to it of f_Λ is actually nondegenerate; so $f_\Gamma = f_{\Lambda|S_\Gamma}$ is a Riemannian structure. In fact, $(\tilde{\Gamma} \circ \pi_F, 1_{FM})$ is a diffeomorphic isometry of (FM, g_Γ) onto (S_Γ, f_Γ). For, if $W \in TS_\Gamma$ then

$$W = 0 \iff T\pi_2(W) = 0$$

and

$$S_\Gamma(W) = f_\Lambda(W) = g_\Gamma(T\pi_2(W), T\pi_2(W)) \,.\square$$

Corollary 9.5.4 *The result is true for reduced subbundles of FM.*

Theorem 9.5.5 *(Conformal stability of connection incompleteness)*
Let (M, g) be a pseudo-Riemannian manifold which is connection incomplete with respect to the Levi-Civita connection Γ^g of g. Let $\{g_\varepsilon\}$ be a 1-parameter family of pseudo-Riemannian metrics with $g_\varepsilon = \varphi_\varepsilon g_0$, $g_0 = g$ and $\varphi_\varepsilon \longrightarrow 1$ as $\varepsilon \longrightarrow 0$. Suppose that $c: [0,1) \longrightarrow FM$ is an inextensible incomplete curve in (FM, g_{Γ^g}) and such that the induced function

$$\varphi_\varepsilon \circ \pi_F \circ c: [0,1) \longrightarrow \mathbf{R}$$

is uniformly convergent to 1 as $\varepsilon \longrightarrow 0$. Then for all sufficiently small $|\varepsilon|$, M is connection incomplete with respect to the Levi-Civita connection Γ^ε of g_ε.

Proof. This follows from the relationship between Levi-Civita connections of conformal metrics. Since the given curve has finite g_{Γ^g}-length it has also finite g_{Γ^ε}-length.$\square$

This result on conformal stability has relevance for physics but it could, of course, be proved directly without reference to systems of connections. However it is easy to prove the following

Theorem 9.5.6 *(Uniform stability of connection-incompleteness)* *Take any $\tilde{\Gamma} \in Sec\,(C_F/M)$ such that M is connection incomplete with respect to the induced connection Γ and let $\chi \in Sec\,(JC_F/C_F)$ be a connection such that $\tilde{\Gamma}$ is χ-horizontal. If $\{\check{\Gamma}_\varepsilon \in Sec\,(C_F/M)\}$ is a 1-parameter family of sections, converging uniformly to Γ for $\varepsilon \longrightarrow 0$ with respect to g_χ, then M is connection-incomplete with respect to Γ_ε for all sufficiently small $|\varepsilon|$.*

In these last two results we have prescriptions for finding open submanifolds $V \subset C_F$ over which incompleteness persists and projects onto connection-incompleteness of FM. Such V can be used in the following general result.

Theorem 9.5.7 *(System stability of connection-incompleteness)*
Suppose $\tilde{\Gamma}_0 \in Sec\,(C_F/M)$ induces Γ_0 with respect to which M is connection-incomplete. As before,

$$S_{\Gamma_0} = (\tilde{\Gamma}_0 \circ \pi_F, 1_{FM})(FM) \,.$$

Let V be an f_Λ-bounded open submanifold of (C_F, f_Λ) such that the f_Γ-boundary of $V \cap S_{\Gamma_0} \subseteq C_F$ contains a point $\bar{x}$ in the f_{Γ_0}-boundary of S_{Γ_0}; such an $\bar{x}$ corresponds to a point x in the g_{Γ_0}-boundary of FM. Let

$$c: [0,1) \longrightarrow FM$$

be a curve in (FM, g_{Γ_0}) *ending at* x. *Now take any* $\tilde{\Gamma} \in Sec\,(C_F/M)$ *with corresponding connection* Γ, *such that*

$$((\tilde{\Gamma} \circ \pi_F, 1_{FM}) \circ c)[0,1) \subseteq V .$$

Then c *is incomplete in* (FM, g_Γ) *and* M *is connection–incomplete with respect to* Γ.

Proof. (Cf. Canarutto and Dodson [8]) Consider the open submanifold $W = \pi_2(V \cap S_{\Gamma_0}) \cap \pi_1(V \cap S_\Gamma)$. It contains c and $x \in \overline{W}$, the g_{Γ_0}–boundary of W. In $\overline{W}$, the open ball of centre x and radius r meets W in an open submanifold W_Γ. Then $\cap_{r>0} W_r$ is empty. Now denote by S_r the g_Γ diameter of W_r and apply the triangle inequality to $c(0) \in W$ and any $b \in W_r$ to obtain

$$\limsup_{r \to 0} \{ dist_{g_\Gamma}(b, c(0)) \mid b \in W_r \} = \liminf_{r \to 0} \{ dist_{g_\Gamma}(b, c(0)) \mid b \in W_r \} .$$

But, if $r > 0$, then $c(t) \in W_r$ for large enough $t \in (0,1)$ so $dist_{g_\Gamma}(c(0), c(t))$ has a definite limit as $t \to 1$.

Therefore $c(\lambda)$ has an endpoint in the g_Γ–boundary of W and hence has finite g_Γ–length.□

The global geometry of Lorentz manifolds is treated by Beem and Ehrlich [4]. They include a discussion of singularities and their stability in spacetime models; further results can be found in [5].

Parametric models in statistical theory

Consider a manifold M as the parameter space of an n–dimensional smooth family of probability distribution functions

$$M = \{ \mathbf{p}_x : \Omega \longrightarrow [0,1] \mid x \in \tilde{M} \}$$

for some fixed event space Ω. Then M is a smooth n–manifold. For each probability distribution function $\mathbf{p}_x$ we have its log likelihood function $\mathbf{l} = \log \mathbf{p}_x$, which depends smoothly on x. We are interested in the situation when at each point $x \in M$ the expected value

$$g_{ij} = E_\Omega \left(\frac{\partial \mathbf{l}}{\partial x^i} \frac{\partial \mathbf{l}}{\partial x^j} \right) \quad \text{(for coordinates } (x^i) \text{ about } x)$$

is a positive definite matrix. In this case it induces a Riemannian metric g on M, called *the expected information metric* for the parametric model, with coordinate expression g_{ij}.

Statistical theorists (Cf. Chentsov [9], Amari [1] or Barndorff–Nielsen et al. [3] for example) have recently studied the so called families of *α-connections* $\{{}^{\alpha}\Gamma \mid \alpha \in \mathbf{R}\}$ on the Riemannian manifold (M, g) of a parametric statistical model, where g is the expected information metric. For these connections the Christoffel symbols are given by

$${}^{\alpha}\Gamma^{k}_{ij} = {}^{0}\Gamma^{k}_{ij} - (1/2)\alpha g^{km} S_{ijm}$$

where ${}^{0}\Gamma$ is the Levi-Civita connection of g and (S_{ijm}) gives the coordinates of the *skewness tensor* of the expectation

$$S_{ijm} = E_{\Omega}\left(\frac{\partial \mathrm{l}}{\partial x^{i}} \frac{\partial \mathrm{l}}{\partial x^{j}} \frac{\partial \mathrm{l}}{\partial x^{m}}\right) .$$

Evidently, the family of α–connections consists of symmetric linear connections on M. In light of earlier results we can incorporate this family in a system of connections; further details are given in Dodson [24].

Theorem 9.5.8 *(Systems of α–connections) Given a parametric statistical model (M, g) the family of α–connections constitutes a system of connections.*

Proof. Consider the system of connections

$$(M \times \breve{\mathbf{R}}) \times FM \longrightarrow JFM \ : (\alpha, b) \mapsto {}^{\alpha}\Gamma(b)$$

where $\breve{\mathbf{R}}$ is $\mathbf{R}$ with the discrete topology and $M \times \breve{\mathbf{R}}$ is then an $\mathbf{R}$–fold covering manifold of M, consisting of smooth copies of M. With this structure, each $\tilde{\Gamma} \in Sec\ (M \times \breve{\mathbf{R}}/M)$ is a constant real function on M, and each such function α defines a unique connection ${}^{\alpha}\Gamma$.□

The construction suggests natural enlargements of the system of α–connections. The following theorem offers two such enlargements.

Theorem 9.5.9 *(Conformal system of variable α–connections) The system of α–connections $\{{}^{\alpha}\Gamma \mid \alpha \in \mathbf{R}\}$ of a parametric statistical model (M, g) is a subsystem of the system*

$$(M \times \mathbf{R}) \times FM \longrightarrow JFM \ : (\alpha, b) \mapsto {}^{\alpha}\Gamma(b) \ (\textit{variable } \alpha\textit{–connections})$$

which itself is a subsystem of

$$(M \times \mathbf{R}^2) \times FM \longrightarrow JFM \ : (\alpha, \varphi, b) \mapsto {}^{\alpha}_{\varphi}\Gamma(b)$$

where ${}^{\alpha}_{\varphi}\Gamma$ is the α–connection arising from the Levi–Civita connection of the metric $e^{\varphi}g$.

Proof. The variable α–connections are obtained from the previous construction by changing $\breve{\mathbf{R}}$ to $\mathbf{R}$ with the usual manifold structure. The second system, consisting of conformal variable α–connections has its members selected by a pair of real functions on M. One function is α as before, the other gives the conformal modification of the metric so

$$ {}^{\alpha}_{\varphi}\Gamma^{k}_{ij} = {}^{0}_{\varphi}\Gamma^{k}_{ij} - (1/2)\alpha e^{-\varphi} g^{km} S_{ijm} $$

and ${}^{0}_{\varphi}\Gamma$ is the Levi–Civita connection $e^{\varphi}g$.

Choosing α =constant gives the required inclusion in the system of variable α–connections; choosing $\varphi = 0$ recovers the latter from our conformal system.□

Remark 9.5.10 The effect of the conformal factor introduced in this system is to allow a weighting of the expected information metric to be applied differentially over the parameter space. The choice of a uniform weighting, φ =constant, recovers the system of variable α–connections; the particular case $\varphi = 0$ recovers the original metric as well.

It remains to be seen what role completeness plays in the differential geometry of parametric statistical models. However, we can easily deduce the following stability results for it, and stability properties are always important in statistical theories.

Theorem 9.5.11 *(Stability of incompleteness in parametric models) If a parametric statistical model (M, g) is geodesically incomplete then the incompleteness persists for a 1–parameter family of metrics conformal to g.*

Proof. This follows from Theorems 8.6.6 and 9.5.5.□

Further discussion of this and other stability results can be found in Dodson [24].

Chapter 10

The Functor J_p^2

Introduction

In the first chapter we constructed a functor J_p^1 by factoring through M the equivalent-gradient classes of real-valued maps on $\mathbf{R}^p$, so yielding the bundle of 1-jets. Now we consider the 2-jets by taking classes having equivalence up to all partial derivatives of second order, so obtaining a functor J_p^2 with similar properties to J_p^1.

The space of all 2-jets at $0 \in \mathbf{R}^n$ of local diffeomorphisms into M is the second order frame bundle F^2M, which is an open dense submanifold of $J_n^2 M$. Now, F^2M is a principal bundle and we study its structure group, connections, fundamental vector fields, G-structures, and lifts of tensor fields. Structural polynomials of $(1,1)$-tensor fields are preserved in diagonal lifts from FM to F^2M, as are Riemannian metrics, symplectic forms and almost Hermitian structures. There is a natural prolongation of G-structures and equivalence of their integrabilities.

10.1 The Bundle $J_p^2 M \longrightarrow M$

Let $f, g \in C^\infty(\mathbf{R}^p)$; we say that f is equivalent to g if

$$\begin{aligned} f(0) &= g(0) \ , \\ \left(\frac{\partial}{\partial t^\alpha}\right)(f) &= \left(\frac{\partial}{\partial t^\alpha}\right)(g) \ , \\ \left(\frac{\partial^2}{\partial t^\alpha \partial t^\beta}\right)(f) &= \left(\frac{\partial^2}{\partial t^\alpha \partial t^\beta}\right)(g) \ , \end{aligned}$$

at $0 \in \mathbf{R}^p$ for all $\alpha, \beta = 1, 2, \ldots, p$.

Let M be an n-dimensional manifold, and consider the set $S_p(M)$ of all maps $\phi \colon \mathbf{R}^p \longrightarrow M$. Let $\phi, \psi \in S_p(M)$; we say that ϕ is equivalent to ψ if $(f \circ \phi)$ is equivalent to $(f \circ \psi)$ for every $f \in C^\infty(M)$. This is again an equivalence relation;

we denote by $j^2(\phi)$ the equivalence class of $\phi \in S_p(M)$. The set of all these equivalence classes will be denoted by J_p^2M.

We define local coordinates on J_p^2M as follows. If (U, x^i) is a local chart in M, we define coordinate functions

$$\{x^i, x^i_\alpha, x^i_{\alpha\beta}, i = 1, 2, \ldots, n,\ \alpha, \beta = 1, 2, \ldots, p\}$$

on J_p^2U by

$$\begin{aligned} x^i(j^2(\phi)) &= x^i(\phi(0)) , \\ x^i_\alpha(j^2(\phi)) &= \frac{\partial(x^i \circ \phi)}{\partial t^\alpha}\Big|_0 , \\ x^i_{\alpha\beta}(j^2(\phi)) &= \frac{\partial^2(x^i \circ \phi)}{\partial t^\alpha \partial t^\beta}\Big|_0 , \end{aligned}$$

where $x^i_{\alpha\beta} = x^i_{\beta\alpha}$.

Then J_p^2M becomes an $(n + pn + (1/2)p(p+1)n)$–dimensional manifold which is called *the tangent bundle of p^2–velocities of M* (see [68]).

Moreover, there are canonical projections

$$\begin{aligned} \pi_1^2 : J_p^2M &\longrightarrow J_p^1M , \\ \pi^2 : J_p^2M &\longrightarrow M , \end{aligned}$$

given by

$$\begin{aligned} \pi_1^2(j^2(\phi)) &= j^1(\phi) , \\ \pi^2(j^2(\phi)) &= \phi(0) . \end{aligned}$$

Then the following diagram

$$\begin{array}{ccc} J_p^2M & \xrightarrow{\pi_1^2} & J_p^1M \\ & {\scriptstyle\pi^2}\searrow \quad \swarrow{\scriptstyle\pi} & \\ & M & \end{array}$$

is commutative.

Functorial properties of J_p^2

It is straightforward to prove that J_p^2 has similar functorial properties to J_p^1.

(1): Let $h: M \longrightarrow N$ be differentiable. Then h induces a canonical differentiable map

$$h^{(2)}: J_p^2M \longrightarrow J_p^2N$$

given by

$$h^{(2)}(j^2(\phi)) = j^2(h \circ \phi) \ , \quad \text{for any } j^2(\phi) \in J^2_p M \ .$$

(2): Let M and N be two differentiable manifolds and $M \times N$ the product manifold. Then $J^2_p(M \times N)$ and $J^2_p M \times J^2_p N$ can be canonically identified.

(3): Let G be a Lie group. Then $J^2_p G$ inherits a canonical Lie group structure. Let $\mu : G \times G \longrightarrow G$ be the group multiplication; then the induced map $\mu^{(2)}$: $J^2_p G \times J^2_p G \longrightarrow J^2_p G$ defines a Lie group multiplication on $J^2_p G$ which is compatible with the manifold structure of $J^2_p G$. Moreover, if G is a Lie subgroup of G' then $J^2_p G$ is a Lie subgroup of $J^2_p G'$.

(4): Let $\rho: M \times G \longrightarrow M$ be a Lie group action of G on M. Then $J^2_p G$ acts on $J^2_p M$ through $\rho^{(2)}: J^2_p M \times J^2_p G \longrightarrow J^2_p M$. If G acts effectively (resp. transitively) on M, then so does $J^2_p G$ on $J^2_p M$.

(5): If $P(M, \pi, G)$ is a principal fibre bundle, $J^2_p P(J^2_p M, \pi^{(2)}, J^2_p G)$ is a principal fibre bundle too. In fact, if $\phi_U: U \times G \longrightarrow \pi^{-1}(U)$ is a local trivialization of P, then

$$\phi^{(2)}_U : J^2_p U \times J^2_p G \longrightarrow J^2_p(\pi^{-1}(U))$$

is a local trivialization of $J^2_p P$.

10.2 The second order frame bundle

Let M be an n-dimensional manifold. Then *the second order frame bundle of* M is the set F^2M of all 2-jets at $0 \in \mathbf{R}^n$ of local diffeomorphisms of open neighbourhoods of $0 \in \mathbf{R}^n$ into M. Then F^2M is an open (dense) submanifold of $J^2_n M$. We shall denote by

$$\pi^2_1 = \pi^2_{1|F^2M}: F^2M \longrightarrow FM \ ,$$

$$\pi^2_M = \pi^2{}_{|F^2M}: F^2M \longrightarrow M$$

the canonical projections. If $\phi: U \subset \mathbf{R}^n \to M$ is a diffeomorphism of a neighbourhood U of $0 \in \mathbf{R}^n$ into an open subset $\phi(U)$ of M, then the 2-jet $j^2(\phi)$ of ϕ at 0 is called a *2-frame*. F^2M is a principal bundle over M with projection π^2_M and with structure group $G^2(n)$, a Lie group which will be described next.

Let $G^2(n)$ be the set of all 2-jets $j^2(g)$ at $0 \in \mathbf{R}^n$, where g is a diffeomorphism from a neighbourhood of 0 in $\mathbf{R}^n$ onto a neighbourhood of 0 in $\mathbf{R}^n$. Then $G^2(n)$ is a group with multiplication defined by the composition of jets, that is,

$$j^2(g) \cdot j^2(g') = j^2(g \circ g') \ .$$

The group $G^2(n)$ acts on F^2M on the right by

$$\begin{array}{ccc} F^2M \times G^2(n) & \longrightarrow & F^2M \\ (j^2(\phi), j^2(g)) & \longmapsto & j^2(\phi \circ g) \ . \end{array}$$

There is a canonical isomorphism

$$G^2(n) \simeq Gl(n, \mathbf{R}) \times S^2(n) \quad ,$$

where $S^2(n)$ is the real vector space of symmetric bilinear forms on $\mathbf{R}^n$, and the multiplication being given by

$$(A, \alpha)(B, \beta) = (AB, \alpha \circ (B, B) + A \circ \beta) \quad .$$

Let $\mathfrak{g}^2(n)$ be the Lie algebra of $G^2(n)$. Then $\mathfrak{g}^2(n)$ can be identified with $gl(n, \mathbf{R}) \oplus S^2(n)$, with a bracket product given by

$$\begin{aligned}[]
&[(A, \alpha), (B, \beta)] \\
&\quad = ([A, B], A \circ \beta - \beta \circ (I, A) - \beta \circ (A, I) - \\
&\qquad\qquad (\beta \circ \alpha - \alpha \circ (I, B) - \alpha \circ (B, I))) \quad ,
\end{aligned}$$

where I is the unit matrix.

With these identifications the adjoint representation of $G^2(n)$ in $\mathfrak{a}(n) = \mathbf{R}^n \oplus gl(n, \mathbf{R})$ is given by

$$Ad^{(2)}(A, \alpha)(v, B) = (Av, \bar{\alpha}(v)A^{-1} + ABA^{-1}) \; ,$$

where $\bar{\alpha}: \mathbf{R}^n \longrightarrow gl(n, \mathbf{R})$ is the linear map defined by

$$\bar{\alpha}(v)(w) = \alpha(v, w) \quad .$$

On the other hand, the adjoint representation of $G^2(n)$ in

$$\mathfrak{g}^{(2)}(n) = gl(n, \mathbf{R}) \oplus S^2(n)$$

is given by

$$\begin{aligned}
Ad(A, \alpha)(B, \beta) = (&ABA^{-1}, \alpha \circ (A^{-1}, BA^{-1}) + \alpha \circ (BA^{-1}, A^{-1}) \\
&- ABA^{-1} \circ \alpha \circ (A^{-1}, A^{-1}) + A \circ \beta \circ (A^{-1}, A^{-1})) \quad .
\end{aligned}$$

Now, let $(A, \alpha) \in \mathfrak{g}^2(n)$. Then (A, α) induces a vector field $\lambda(A, \alpha)$ on F^2M, called the *fundamental vector field* corresponding to (A, α). Then the vertical subspace (with respect to $\pi_M^2: F^2M \longrightarrow M$) at any point $p \in F^2M$ can be decomposed as follows:

$$V_p(F^2M) = \lambda(gl(n, \mathbf{R}))_p \oplus \lambda(S^2(n))_p \quad .$$

The *canonical form* $\theta^{(2)}$ *of* F^2M is defined as follows. Let

$$p = j^2(\varphi) \in F^2M \;\text{ and }\; \bar{p} = \pi_1^2(p) = j^1(\varphi) \in FM \quad ,$$

where φ is a local diffeomorphism of a neighbourhood U of $0 \in \mathbf{R}^n$ into an open subset $\varphi(U)$ of M. Then, the induced map

$$\varphi^1 : FU \longrightarrow F(\varphi(U))$$

is a principal bundle isomorphism. Let $e = j^1(1_{\mathbf{R}^n})$; then

$$\varphi^1(e) = \varphi^1(j^1(1_{\mathbf{R}^n})) = j^1(\varphi) = \bar{p} \ .$$

Therefore, we obtain a linear isomorphism

$$(\varphi^1)_*(e) : T_e(FU) \longrightarrow T_{\bar{p}}(F(\varphi(U))) \simeq T_{\bar{p}}(FM) \ .$$

Since $T_e(FU) \equiv T_e(F\mathbf{R}^n)$ can be identified with $\mathbf{R}^n \oplus gl(n, \mathbf{R})$, then $(\varphi^1)_*(e)$ defines a linear isomorphism

$$(\varphi^1)_*(e) : \mathbf{R}^n \oplus gl(n, \mathbf{R}) \longrightarrow T_{\bar{p}}(FM) \ .$$

Now, define

$$(\theta^{(2)})_p(X) = ((\varphi^1)_*(e))^{-1}((\pi_1^2)_* X)) \ , \quad X \in T_p(F^2M) \ .$$

It is not hard to prove that $\theta^{(2)}$ is an $(\mathbf{R}^n \oplus gl(n, \mathbf{R}))$-valued form of type $(Ad^{(2)}, G^2(n))$ satisfying

$$\theta^{(2)}(\lambda(A, \alpha)) = A \ .$$

Let $\theta^{(2)} = \theta_{-1} + \theta_0$ be the canonical decomposition of $\theta^{(2)}$, so θ_{-1} is an $\mathbf{R}^n$-valued 1-form and θ_0 a $gl(n, \mathbf{R})$-valued 1-form on F^2M. Then

$$\begin{aligned} \theta_{-1}(\lambda(A, \alpha)) &= 0 \ , \\ \theta_0(\lambda(A, \alpha)) &= A \ . \end{aligned}$$

A simple computation establishes the identity

$$\theta_{-1} = (\pi_1^2)^* \theta \ ,$$

where θ denotes the canonical 1-form on FM.

If $\{e_i\}$ and $\{E_i^j\}$ denote the canonical bases of $\mathbf{R}^n$ and $gl(n, \mathbf{R})$, respectively, then we put

$$\theta_{-1} = \theta^i e_i \ , \quad \theta_0 = \theta^i_j E^j_i \ ,$$

with

$$\left\{ \begin{aligned} \theta^i &= y^i_k dx^k \ , \\ \theta^i_j &= y^i_k (dx^k_j - x^k_{hj} y^h_l dx^l) \ , \end{aligned} \right. \tag{10.1}$$

where $(y^i_j) = (x^i_j)^{-1}$. From (10.1) we easily obtain the following *structure equation*:

$$d\theta^i = -\theta^i_k \wedge \theta^k \ .$$

10.3 Second order connections

A connection Γ in F^2M will be called a *second order connection on* M.

Let ω be the connection form of a second order connection Γ; ω is a 1-form on F^2M of type $(Ad, G^2(n))$. Then, ω can be decomposed as

$$\omega = \omega_0 + \omega_1 \quad ,$$

where ω_0 is the $gl(n, \mathbf{R})$-component and ω_1 the $S^2(n)$-component of ω. Since

$$\omega(\lambda(A, \alpha)) = (A, \alpha) \quad ,$$

then

$$\omega_0(\lambda(A, \alpha)) = A \quad , \quad \omega_1(\lambda(A, \alpha)) = \alpha \quad .$$

Let $\Theta = D\theta^{(2)}$, $\Omega = D\omega$ the torsion and the curvature forms of Γ, respectively. Then Θ (resp. Ω) is a tensorial 2-form of type $(Ad^{(2)}, G^2(n))$ (resp. of type $(Ad, G^2(n))$) on F^2M. Thus Θ and Ω can be decomposed as follows:

$$\begin{aligned} \Theta &= \Theta_{-1} + \Theta_0 \quad , \\ \Omega &= \Omega_0 + \Omega_1 \quad . \end{aligned}$$

A simple computation shows:

$$\begin{aligned} \Theta_{-1} &= D\theta_{-1} \quad , \quad \Theta_0 = D\theta_0 \quad , \\ \Omega_0 &= D\omega_0 \quad , \quad \Omega_1 = D\omega_1 \quad . \end{aligned}$$

For any 2-frame $p \in F^2M$, $(\theta_{-1})_p$ gives a linear isomorphism of the horizontal subspace H_p at p onto $\mathbf{R}^n$. Thus, to each $\xi \in \mathbf{R}^n$ we associate a horizontal vector field $C(\xi)$ on F^2M as follows. For each $p \in F^2M$, $C(\xi)_p$ is the unique horizontal vector at p such that

$$(\theta_{-1})_p(C(\xi)_p) = \xi \quad ;$$

$C(\xi)$ is called *the standard horizontal vector field on* F^2M corresponding to ξ.

Proposition 10.3.1 *The standard horizontal vector fields satisfy the following properties:*
(1): $R_{(A,\alpha)}C(\xi) = C(A^{-1}\xi)$ *for all* $(A, \alpha) \in G^2(n)$ *and* $\xi \in \mathbf{R}^n$.
In particular, $R_\alpha C(\xi) = 0$ *for all* $\alpha \in S^2(n)$.
(2): *If* $\xi \neq 0$, *then* $C(\xi)$ *never vanishes.*

Proof. **(1)** follows from the fact that if X is horizontal at p then $R_{(A,\alpha)}X$ is horizontal at $p(A, \alpha)$ and θ is of type $(Ad^{(2)}, G^2(n))$. To prove **(2)**, assume $C(\xi)_p = 0$ at some $p \in F^2M$; then $0 = (\theta_{-1})_p C(\xi)_p = \xi$.□

Proposition 10.3.2

$$[\lambda(A,\alpha), C(\xi)] = C(A\xi) \ .$$

In particular

$$\begin{aligned} [\lambda A, C(\xi)] &= C(A\xi) \ , \\ [\lambda\alpha, C(\xi)] &= 0 \ , \end{aligned}$$

for all $A \in gl(n, \mathbf{R})$, $\alpha \in S^2(n)$ *and* $\xi \in \mathbf{R}^n$.

Proof. Let (a_t, α_t) be the 1-parameter subgroup of $G^2(n)$ generated by (A, α), where $a_t = exp\, tA$. Then

$$\begin{aligned} [\lambda(A,\alpha), c(\xi)] &= \lim_{t\to 0}\tfrac{1}{t}[C(\xi) - R_{(a_t,\alpha_t)}C(\xi)] \\ &= \lim_{t\to 0}\tfrac{1}{t}[C(\xi) - C(a_t^{-1}\xi)] \ . \end{aligned}$$

Now, since $\xi \mapsto C(\xi)_p$ is a linear isomorphism of $\mathbf{R}^n$ onto H_p, we have

$$\lim_{t\to 0}\frac{1}{t}\left[C(\xi) - C(a_t^{-1}\xi)\right] = C\left[\lim_{t\to 0}\frac{1}{t}(\xi - a_t^{-1}\xi)\right] = C(A\xi) \quad .\square$$

The structure equation of Γ is

$$d\omega = \Omega - (1/2)[\omega, \omega] \ ,$$

or, equivalently,

$$(10.2) \qquad \left\{ \begin{aligned} d\omega_0 &= \Omega_0 - (1/2)[\omega_0, \omega_0] \ , \\ d\omega_1 &= \Omega_1 - (1/2)\{[\omega_0, \omega_1] + [\omega_1, \omega_0]\} \ . \end{aligned} \right.$$

Let $\{E_i^{jk}\}$, with $E_i^{jk} = E_i^{kj}$, be the canonical basis of $S^2(n)$. Then

$$\begin{aligned} \omega_0 &= \omega_j^i E_i^j \quad , \quad \omega_1 = \omega_{jk}^i E_i^{jk} \ , \\ \Omega_0 &= \Omega_j^i E_i^j \quad , \quad \Omega_1 = \Omega_{jk}^i E_i^{jk} \ , \end{aligned}$$

where $\omega_{jk}^i = \omega_{kj}^i$, $\Omega_{jk}^i = \Omega_{kj}^i$, and (10.2) can be equivalently written as

$$\begin{aligned} d\omega_j^i &= \Omega_j^i - \omega_k^i \wedge \omega_j^k \ , \\ d\omega_{jk}^i &= \Omega_{jk}^i - \omega_l^i \wedge \omega_{jk}^l + \omega_k^l \wedge \omega_{jl}^i + \omega_j^l \wedge \omega_{lk}^i \ . \end{aligned}$$

Next, let σ be the local section of F^2M given by $\sigma(x) = (x^i, I, 0)$ over (U, x^i). We define functions Γ_{jk}^i, Γ_{jkl}^i on U, with $\Gamma_{jkl}^i = \Gamma_{jlk}^i$, by

$$\begin{aligned} \sigma^*\omega_0 &= (\Gamma_{jk}^i dx^j)E_i^k \ , \\ \sigma^*\omega_1 &= (\Gamma_{jkl}^i dx^j)E_i^{kl} \ . \end{aligned}$$

The functions Γ^i_{jk}, Γ^i_{jkl} are called the *components of* Γ *with respect to* (U, x^i).

A straightforward computation shows that

$$\begin{aligned}
\omega^i_j &= y^i_k(dx^k_j + \Gamma^k_{ml}x^l_j\, dx^m) \ , \\
\omega^i_{jk} &= \Big\{-\Gamma^r_{ms}(x^s_k y^c_r y^i_u x^u_{jc} + x^s_j y^l_r y^i_u x^u_{lk} - x^s_{jk} y^i_r) \\
&\quad + \Gamma^r_{msl} x^s_j x^l_k y^i_r\Big\}\, dx^m \\
&\quad - y^l_r y^i_s x^s_{lk}\, dx^r_j - y^l_r y^i_s x^s_{jl}\, dx^r_k + y^i_s\, dx^s_{jk} \ .
\end{aligned}$$

On the other hand, if we put

$$\begin{aligned}
\sigma^*\Omega_0 &= (\sigma^*\Omega^i_j)\, E^j_i &&= (1/2)(R^i_{jkl}\, dx^k \wedge dx^l)E^j_i \ , \\
\sigma^*\Omega_1 &= (\sigma^*\Omega^i_{jk})\, E^{jk}_i &&= (1/2)(R^i_{jkrl}\, dx^r \wedge dx^l)E^{jk}_i \ ,
\end{aligned}$$

then we obtain

$$\begin{aligned}
R^i_{jkl} &= \partial_k\Gamma^i_{lj} - \partial_l\Gamma^i_{kj} + \Gamma^m_{lj}\Gamma^i_{km} - \Gamma^m_{kj}\Gamma^i_{lm} \ , \\
R^i_{jkrl} &= \partial_r\Gamma^i_{ljk} - \partial_l\Gamma^i_{rjk} + \Gamma^s_{lk}\Gamma^i_{rjs} - \Gamma^s_{rk}\Gamma^i_{ljs} \\
&\quad + \Gamma^s_{lj}\Gamma^i_{rsk} - \Gamma^s_{rj}\Gamma^i_{lsk} + \Gamma^i_{rs}\Gamma^s_{ljk} - \Gamma^i_{ls}\Gamma^s_{rjk} \ .
\end{aligned}$$

Now, let $\xi = \xi^i\, e_i \in \mathbf{R}^n$. Then

$$C(\xi) = x^i_m\xi^m\left(\frac{\partial}{\partial x^i} - \Gamma^k_{il}x^l_j\,\frac{\partial}{\partial x^k_j}(\Gamma^s_{ir}x^r_{jk} + \Gamma^s_{irl}x^r_j x^l_k)\frac{\partial}{\partial x^s_{jk}}\right) \ .$$

Observe that the n global vector fields $C(e_i)$ span the horizontal distribution H_Γ on F^2M. Therefore, the $n + n^2 + (n^2(n+1)/2)$ global vector fields $\{C(e_i), \lambda E^i_j, \lambda E^{jk}_i, j \leq k\}$ define an absolute parallelism on F^2M and are dual to θ^i, ω^i_j, ω^i_{jk}. Moreover, the local expressions of λE^i_j and λE^{jk}_i on F^2U are

$$\begin{aligned}
\lambda E^i_j &= x^r_i\,\frac{\partial}{\partial x^r_j} + x^r_{is}\,\frac{\partial}{\partial x^r_{js}} + x^r_{si}\,\frac{\partial}{\partial x^r_{sj}} \ , \\
\lambda E^{jk}_i &= x^r_i\,\frac{\partial}{\partial x^r_{jk}} \ .
\end{aligned}$$

Thus, the horizontal distribution H_Γ is spanned by the local vector fields

$$D_i = \frac{\partial}{\partial x^i} - \Gamma^k_{il}x^l_r\,\frac{\partial}{\partial x^k_r} - \left(\Gamma^s_{im}x^m_{rk} + \Gamma^s_{iml}x^m_r x^l_k\right)\frac{\partial}{\partial x^s_{rk}} \ ,$$

and the vertical distribution V is spanned by the local vector fields

$$D^i_j = \frac{\partial}{\partial x^i_j} + y^r_j x^m_{rs} \frac{\partial}{\partial x^m_{is}} + y^r_j x^m_{sr} \frac{\partial}{\partial x^m_{si}} ,$$
$$D^i_{jk} = \frac{\partial}{\partial x^i_{jk}} ,$$

when we restrict to F^2U.

The frame $\{D_i, D^i_k, D^i_{kj}\}$ is adapted to the almost product structure (H_Γ, V) and we shall call it *the adapted frame on* F^2U. The local 1–forms η^i, η^i_k, η^i_{jk} on F^2U dual to $\{D_i, D^i_k, D^i_{kj}\}$ are given by

$$\eta^i = dx^i ,$$
$$\eta^i_j = \Gamma^i_{rs} x^s_j \, dx^r + dx^i_j ,$$
$$\eta^i_{jk} = \Big\{(\Gamma^i_{lr} x^r_{jk} + \Gamma^i_{lrs} x^r_j x^s_k) - y^s_m \Gamma^m_{lr}(x^r_j x^i_{sk} + x^r_k x^i_{js})\Big\}\, dx^l - y^r_l(\delta^{sj} x^i_{rk} + \delta^{sk} x^i_{jr})\, dx^l_s + dx^i_{jk} ,$$

and $\{\eta^i, \eta^i_j, \eta^i_{jk}\}$ will be called *the adapted coframe on* F^2U.

Let Γ be a connection of order 2 on M with connection form ω. The canonical projection $\pi^2_1 : F^2M \longrightarrow FM$ is a principal bundle morphism which induces the canonical projection

$$G^2(n) = Gl(n, \mathbf{R}) \times S^2(n) \longrightarrow Gl(n, \mathbf{R})$$

between the corresponding structure groups. Then π^2_1 sends the connection Γ onto a connection $\widetilde{\Gamma}$ on FM, that is, $\widetilde{\Gamma}$ is a linear connection on M. In fact the horizontal subspace $\widetilde{H}_{\tilde{p}}$ at $\tilde{p} \in FM$ is

$$\widetilde{H}_{\tilde{p}} = (\pi^2_1)_* H_p ,$$

where H_p is the horizontal subspace of Γ at p and $\pi^2_1(p) = \tilde{p}$. Moreover,

$$(\pi^2_1)_* \lambda(A, \alpha) = \lambda A , \quad (\pi^2_1)_* C(\xi) = B(\xi) ,$$

where λA (resp. $B(\xi)$) is the fundamental (resp. standard horizontal) vector field on FM corresponding to A (resp. to ξ). Therefore, if $\tilde{\theta}$ is the canonical form on FM,

$$(\pi^2_1)^* \tilde{\theta} = \theta_{-1} .$$

Thus, if $\widetilde{\Theta}$ is the torsion form of $\widetilde{\Gamma}$,

$$(\pi^2_1)^* \widetilde{\Theta} = \Theta_{-1} .$$

On the other hand, if $\tilde{\omega}$ (resp. $\tilde{\Omega}$) denotes the connection form (resp. the curvature form) of $\tilde{\Gamma}$,

$$(\pi_1^2)^*\tilde{\omega} = \omega_0 \quad , \quad (\pi_1^2)^*\tilde{\Omega} = \Omega_0 \quad .$$

If $\tilde{\sigma}$ is a local section of FM over U given by $\tilde{\sigma}(x) = (x^i, I)$, then the components $\tilde{\Gamma}^i_{jk}$ of $\tilde{\Gamma}$ are given by

$$\tilde{\sigma}^*\tilde{\omega} = (\tilde{\Gamma}^i_{jk}\, dx^j)E^i_k \quad .$$

Then,

$$\begin{aligned} \tilde{\Gamma}^i_{jk} &= \Gamma^i_{jk} \quad , \\ \tilde{R}^i_{jkl} &= R^i_{jkl} \quad , \end{aligned}$$

$\tilde{R}^i_{jkl}$ being the components of the curvature tensor of $\tilde{\Gamma}$.

10.4 Geodesics of second order

We shall give an interpretation of the geodesics of second order in terms of standard horizontal vector fields similar to that of the case of linear connections.

Let T^2M be the tangent bundle of second order of M; that is, T^2M is the set of 2-jets at $0 \in \mathbf{R}$ of curves in M. Obviously, $T^2M = J_1^2M$. Then there exist canonical projections

$$\begin{aligned} \pi^2 &: T^2M \longrightarrow M \quad , \\ \pi_1^2 &: T^2M \longrightarrow TM \quad . \end{aligned}$$

It is well known that T^2M (resp. TM) is an associate fibre bundle with fibre $\mathbf{R}^{2n}$ (resp. $\mathbf{R}^n$) to F^2M (resp. to FM). Then, if Γ is a second order connection on M it defines a connection Γ' in T^2M, and the induced linear connection $\tilde{\Gamma}$ defines a connection $\tilde{\Gamma}'$ in TM. Thus the horizontal subspaces of Γ' are mapped into the horizontal subspaces of $\tilde{\Gamma}'$ by $\pi_1^2 : T^2M \to TM$.

Let σ be a curve in M, and denote by $j^2\sigma$ the canonical prolongation of σ to T^2M.

Definition 10.4.1 [33] *σ is called a geodesic of second order with respect to Γ if $j^2\sigma$ is horizontal with respect to Γ'.*

The following proposition is easy from this definition:

Proposition 10.4.2 *If σ is a geodesic of second order in M with respect to Γ, then σ is a geodesic with respect to $\tilde{\Gamma}$.*

Definition 10.4.3 [33] *A second order connection Γ on M is called admissible if for each point $x \in M$ and each tangent vector $X \in T_xM$ there is a geodesic of second order σ such that $\sigma(0) = x$ and $\dot{\sigma}(0) = X$.*

If such a geodesic exists, then it is unique, since σ is a geodesic with respect to $\tilde{\Gamma}$.

Proposition 10.4.4 *If Γ is an admissible second order connection on M then every geodesic with respect to $\tilde{\Gamma}$ is a geodesic of second order with respect to Γ.*

Proof. Let σ be a geodesic in M with respect to $\tilde{\Gamma}$. We set $x = \sigma(0)$, $X = \dot{\sigma}(0) \in T_xM$. Since Γ is admissible, there exists on M a unique geodesic of second order γ with respect to Γ such that $\gamma(0) = x$, $\dot{\gamma}(0) = X$. But γ is also a geodesic with respect to $\tilde{\Gamma}$. From the uniqueness of γ it follows $\gamma = \sigma$.□

It is known that if Γ is a linear connection on M, then the geodesics of Γ are the projections onto M of integral curves of standard horizontal vector fields on FM. The same property holds for admissible second order connections.

Theorem 10.4.5 [51] *Let Γ be a second order connection on M. Then every geodesic of second order is the projection onto M of an integral curve of some standard horizontal vector field on F^2M. Conversely, if Γ is admissible then the projection onto M of any integral curve of a standard horizontal vector field on F^2M is a geodesic of second order.*

Proof. Let $\sigma(t)$ be a geodesic of second order with respect to Γ. We can suppose that $\sigma(t)$ is defined in some open interval containing $0 \in \mathbf{R}$. Let $u_0 \in F^2M$ be a 2-frame at $\sigma(0)$. We set $p_0 = \pi_1^2(u_0)$ and $\xi = p_0^{-1}(\dot{\sigma}(0))$. Let $\tau(t)$ (resp $\gamma(t)$) be the horizontal lift of $\sigma(t)$ through p_0 (resp. u_0) with respect to the induced linear connection $\tilde{\Gamma}$ (resp. with respect to Γ). Then $\pi_1^2 \circ \gamma = \tau$. Since $\sigma(t)$ is a geodesic of $\tilde{\Gamma}$, we have $\dot{\sigma}(t) = \tau(t)(\xi)$. Moreover,

$$\theta(\dot{\tau}(t)) = \tau(t)^{-1}(\pi(\dot{\tau}(t))) = (\tau(t))^{-1}(\dot{\sigma}(t)) = \xi \quad .$$

Thus $\tau(t)$ is an integral curve of $B(\xi)$. Therefore $\gamma(t)$ is an integral curve of $C(\xi)$, since

$$\theta_{-1}(\dot{\gamma}(t)) = ((\pi_1^2)^*\theta)(\dot{\gamma}(t)) = \theta((\pi_1^2)_*\dot{\gamma}(t)) = \theta(\dot{\tau}(t)) = \xi \quad ,$$

which implies $\dot{\gamma}(t) = C(\xi)_{\sigma(t)}$.

Conversely, let Γ be an admissible second order connection on M and $C(\xi)$ a standard horizontal vector field, $\xi \in \mathbf{R}^n$. Let $\gamma(t)$ be an integral curve of $C(\xi)$. We set $\tau(t) = \pi_1^2(\gamma(t))$ and then $\tau(t)$ is an integral curve of $B(\xi)$ on FM with respect to $\tilde{\Gamma}$. Thus $\sigma(t) = \pi^2(\gamma(t))$ is a geodesic on M with respect to $\tilde{\Gamma}$ and, from Proposition 10.4.4, a geodesic of second order with respect to Γ.□

Corollary 10.4.6 *Let Γ be an admissible second order connection on M. Then a curve σ in M is a geodesic of second order with respect to Γ if and only if σ is the projection onto M of an integral curve of some standard horizontal vector field on F^2M.*

10.5 G–structures on F^2M

A procedure similar to that described in Chapter 8 allows us to define a great variety of G–structures over F^2M using a second order connection.

Let Γ be a second order connection on M with connection form ω; hence the family of vector fields $\{C(e_i), \lambda E_j^i, \lambda E_i^{jk}, j \leq k\}$ defines an absolute parallelism on F^2M. Then there is an associated trivialization of the frame bundle FF^2M given by

$$\tau: F^2M \times Gl(N, \mathbf{R}) \longrightarrow FF^2M \ ,$$

where $N = n + n^2 + (n^2(n+1)/2)$, by setting

$$\tau(p, A) = \tilde{p}_0 A \ , \quad p \in F^2M \ , \ A \in Gl(N, \mathbf{R}) \ ,$$

$\tilde{p}_0$ being the linear frame of $T_p(F^2M)$ given by

$$\tilde{p}_0 = \{C(e_i)_p, (\lambda E_j^i)_p, (\lambda E_i^{jk})_p, j \leq k\} \ .$$

Let $G \subset Gl(N, \mathbf{R})$ be a Lie subgroup of $Gl(N, \mathbf{R})$. Then the principal bundle

$$P_G = \tau(F^2M \times G)$$

defines a G–structure on F^2M which will be called the *ω-associated G–structure on F^2M.*

Now let $\tilde{\Gamma}$ be the linear connection induced by Γ. Then the family of vector fields $\{Be_i, \lambda E_j^i\}$ defines an absolute parallelism on FM (c.f. 8.1) which admits a trivialization of FFM

$$\tilde{\tau}: FM \times Gl(n + n^2, \mathbf{R}) \longrightarrow FFM \ ,$$

in such a way that the following diagram is commutative:

$$\begin{array}{ccc} F^2M \times Gl(N, \mathbf{R}) & \xrightarrow{\tau} & FF^2M \\ \downarrow & & \downarrow \\ FM \times Gl(n + n^2, \mathbf{R}) & \xrightarrow{\tilde{\tau}} & FFM \end{array}$$

where the vertical arrows are the natural maps.

So, if $\tilde{G} \subset Gl(n+n^2, \mathbf{R})$ is the projection of $G \subset Gl(N)$, then the G–structure $P_G = \tau(F^2M \times G)$ projects onto the $\tilde{G}$–structure $P_{\tilde{G}} = \tilde{\tau}(FM \times \tilde{G})$ on FM. ($P_{\tilde{G}}$ is the $\tilde{\omega}$–associated $\tilde{G}$–structure on FM defined in 8.1.)

10.6 Vector fields on F^2M

Next, we introduce a wide class of vector fields on F^2M and obtain some identities which will be useful later.

First, let $P(M,\pi,G)$ be an arbitrary principal bundle over M. If $\mathfrak{g}$ denotes the Lie algebra of G, then for any function $f: P \to \mathfrak{g}$ we can define the vertical vector field λf given by

$$(\lambda f)(p) = (\lambda(f(p)))_p \quad , \quad p \in P \ , \tag{10.3}$$

and λf will be called *the fundamental vector field corresponding to f*.

On the other hand, if Γ is a connection in P and X^H is the horizontal lift of a vector field X on M to P, then straightforward computations show

$$[\lambda f, \lambda g] = \lambda[f,g] + \lambda(\lambda f(g)) - \lambda(\lambda g(f)) \ , \tag{10.4}$$

$$[X^H, \lambda g] = \lambda(X^H g) \ , \tag{10.5}$$

$$[X^H, Y^H] = [X,Y]^H - 2\lambda\Omega(X^H, Y^H) \ , \tag{10.6}$$

for all vector fields X, Y on M and all functions $f, g: P \to \mathfrak{g}$.

Remark 10.6.1 If f and g are the constant functions A and B respectively, $A, B \in \mathfrak{g}$, then (10.4) and (10.5) are reduced to

$$\begin{aligned} [\lambda A, \lambda B] &= \lambda[A,B] \ , \\ [X^H, \lambda B] &= 0 \ . \end{aligned}$$

Let us return to the second order frame bundle F^2M, and assume that there is given a second order connection Γ on M. Let X be a vector field on M with local expression in U

$$X = X^i \frac{\partial}{\partial x^i} \ .$$

Then the local expression of X^H on F^2U is

$$X^H = X^i \left(\frac{\partial}{\partial x^i} - \Gamma^r_{il} x^l_j \frac{\partial}{\partial x^r_j} - (\Gamma^r_{il} x^l_{jk} + \Gamma^r_{ilm} x^l_j x^m_k) \frac{\partial}{\partial x^r_{jk}} \right) \ , \tag{10.7}$$

that is,

$$X^H = X^i D_i$$

with respect to the adapted frame.

Remark 10.6.2 If $X^{\widetilde{H}}$ is the horizontal lift of X to FM with respect to the induced linear connection $\widetilde{\Gamma}$, then

$$(\pi_1^2)_* X^H = X^{\widetilde{H}}$$

Now, let F be a $(1,1)$–tensor field on M. We define a function

$$F^0 \colon F^2M \longrightarrow gl(n, \mathbf{R})$$

as follows. For any $p \in F^2M$, let $\tilde{p} = \pi_1^2(p)$ be the linear frame at $x = \pi^2(p)$. If $\widetilde{F}^0(\tilde{p})$ is the matrix representation of F_x with respect to $\tilde{p}$, we put

$$F^0(p) = \widetilde{F}^0(\tilde{p}) \ .$$

If F has local components $\{F_i^h\}$ in U, then in F^2U

$$F^0 = F_s^h x_j^s y_h^j$$

and the corresponding fundamental vector field is given by

$$\lambda F^0 = F_s^r x_j^s \frac{\partial}{\partial x_j^r} + F_s^h x_j^s y_h^i x_{il}^r \frac{\partial}{\partial x_{jl}^r} + F_s^h x_j^s y_h^i x_{li}^r \frac{\partial}{\partial x_{lj}^r} \ . \tag{10.8}$$

On the other hand, if $A \in gl(n, \mathbf{R})$ and $\alpha \in S^2(n)$, we can define functions

$$\begin{aligned} F^0A \colon F^2M &\longrightarrow gl(n, \mathbf{R}) \ , \\ F^0\alpha \colon F^2M &\longrightarrow S^2(n) \ , \end{aligned}$$

by

$$\begin{aligned} (F^0A)(p) &= F^0(p)A \ , \\ (F^0\alpha)(p) &= F^0(p) \circ \alpha \ , p \in F^2M \ . \end{aligned}$$

Thus, the corresponding fundamental vector fields $\lambda(F^0A)$ and $\lambda(F^0\alpha)$ are locally expressed by

$$\begin{aligned} \lambda(F^0A) &= F_j^h x_s^j A_t^s \frac{\partial}{\partial x_t^h} + F_j^h x_s^j y_h^r A_t^s x_{rk}^i \frac{\partial}{\partial x_{tk}^i} \\ &\quad + F_j^h x_s^j y_h^r A_t^s x_{kr}^i \frac{\partial}{\partial x_{kt}^i} \ , \end{aligned} \tag{10.9}$$

$$\lambda(F^0\alpha) = F_j^h x_s^j \alpha_{mr}^s \frac{\partial}{\partial x_{mr}^h} \ , \tag{10.10}$$

where

$$A = A^i_j E^j_i \quad , \quad \alpha = \alpha^i_{jk} E^{jk}_i \quad , \quad \alpha^i_{jk} = \alpha^i_{kj} \quad .$$

The following formulas can be obtained by direct computation from (10.4)–(10.10):

$$(10.11) \quad \begin{cases} [X^H, \lambda F^0] = \lambda(\nabla_X F)^0 , \\ [X^H, \lambda(F^0 A)] = \lambda((\nabla_X F)^0 A) , \\ [X^H, \lambda(F^0 \alpha)] = \lambda((\nabla_X F)^0 \alpha) , \\ [\lambda(F^0 A), \lambda B] = \lambda(F^0[A, B]) , \\ [\lambda(F^0 A), \lambda\alpha] = \lambda([F^0 A, \alpha] = \lambda\{F^0(A \circ \alpha) \\ \qquad - \alpha \circ (I, F^0 A) - \alpha \circ (F^0 A, I)\} , \\ [\lambda(F^0 A), \lambda(F^0 B)] = \lambda((F^2)^0[A, B]) , \\ [\lambda(F^0 A), \lambda(F^0 \beta)] = \lambda\{(F^2)^0(A \circ \beta) - (F^0\beta) \circ (I, F^0 A) \\ \qquad - (F^0\beta) \circ (F^0 A, I)\} , \\ [\lambda(F^0 \alpha), \lambda B] = \lambda(F^0[\alpha, B]) = -\lambda\{F^0(B \circ \alpha) \\ \qquad - F^0(\alpha \circ (I, B)) - F^0(\alpha \circ (B, I))\} , \\ [\lambda(F^0 \alpha), \lambda\beta] = [\lambda(F^0\alpha), \lambda(F^0\beta)] = 0 , \end{cases}$$

for all vector fields X and $(1,1)$–tensor fields F on M, and all A, $B \in gl(n, \mathbf{R})$, α, $\beta \in S^2(n)$, and where ∇ denotes the covariant derivative of the induced linear connection Γ.

Finally, (10.6) can be rewritten as follows:

$$[X^H, Y^H] = [X, Y]^H - 2\lambda\Omega_0(X^H, Y^H) - 2\lambda\Omega_1(X^H, Y^H) \quad .$$

Since $2\lambda\Omega_0(X^H, Y^H) = \lambda(R(X, Y)^0)$, we obtain

$$(10.12) \qquad [X^H, Y^H] = [X, Y]^H - \lambda(R(X, Y)^0) - 2\lambda\Omega_1(X^H, Y^H) \ ,$$

for all vector fields X, Y on M.

10.7 Diagonal lifts of tensor fields

Algebraic preliminaries

Let $u \in (\mathbf{R}^n)^*$; then we define $u' \in gl(n, \mathbf{R})^*$ and $u'' \in S^2(n)^*$ as follows:

$$u'(A) = \sum_{j=1}^{n} u(A_j) \quad , \quad A \in gl(n, \mathbf{R}) \ ,$$

$$u''(\alpha) = \sum_{j,k=1}^{n} u(\alpha_{jk}) \quad , \quad \alpha \in S^2(n) \ ,$$

where A_j (resp. α_{jk}) denotes the jth–column (resp. (j,k)–column) of A (resp. of α).

Let $u \in Hom(\mathbf{R}^n, \mathbf{R}^n)$; then u induces $u' \in Hom(gl(n,\mathbf{R}), gl(n,\mathbf{R}))$ and $u'' \in Hom(S^2(n), S^2(n))$ given as follows:

$$\begin{aligned} u'(A) &= u \circ A \quad , \quad A \in gl(n,\mathbf{R}) \ , \\ u''(\alpha) &= u \circ \alpha \quad , \quad \alpha \in S^2(n) \ . \end{aligned}$$

If $rank\, u = r$, then $rank\, u' = rn$ and $rank\, u'' = rn(n+1)/2$.

Let $u \in \otimes_s(\mathbf{R}^n)^*$, $s \geq 1$; then there exist elements $u' \in \otimes_s(gl(n,\mathbf{R}))^*$ and $u'' \in \otimes_s(S^2(n))^*$ defined by

$$\begin{aligned} u'(A_1, \ldots, A_s) &= \sum_{j=1}^{n} u((A_1)_j, \ldots, (A_s)_j) \ , \quad A_1, \ldots, A_s \in gl(n,\mathbf{R}) \ , \\ u''(\alpha_1, \ldots, \alpha_s) &= \sum_{j,k=1}^{n} u((\alpha_1)_{jk}, \ldots, (\alpha_s)_{jk})) \ , \quad \alpha_1, \ldots, \alpha_s \in S^2(n) \ . \end{aligned}$$

When $s = 2$ we have

$$\begin{aligned} u'(A,B) &= \sum_{j=1}^{n} u(A_j, B_j) \quad , \quad A, B \in gl(n,\mathbf{R}) \ , \\ u''(\alpha,\beta) &= \sum_{j,k=1}^{n} u(\alpha_{jk}, \beta_{jk}) \quad , \quad \alpha, \beta \in S^2(n) \ . \end{aligned}$$

It is easy to prove that if u is symmetric (resp. skew–symmetric) then u' and u'' are symmetric (resp. skew–symmetric) too. Moreover, if $rank\, u = r$ then $rank\, u' = rn$ and $rank\, u'' = rn(n+1)/2$.

Let $u \in \mathbf{R}^n \otimes (\otimes_s(\mathbf{R}^n)^*)$, $s \geq 1$; then there exist elements

$$u' \in gl(n,\mathbf{R}) \otimes (\otimes_s(gl(m,\mathbf{R}))^*) \ \text{ and } \ u'' \in S^2(n) \otimes (\otimes_s(S^2(n))^*)$$

such that the jth–column (resp. the (j,k)–column) of $u'(A_1, \ldots, A_s)$ (resp. of $u''(\alpha_1, \ldots, \alpha_s)$) is $u((A_1)_j, \ldots, (A_s)_j)$ (resp. $u((\alpha_1)_{jk}, \ldots, (\alpha_s)_{jk})$).

Diagonal lifts of 1–forms

Let τ be a 1–form on M. For each $p \in F^2M$, we put

$$u_p = \tau_x \circ \pi_1^2(p) \ ,$$

where $x = \pi^2(p)$. Then the *diagonal lift* τ^D *of* τ *to* F^2M is the 1–form given by

$$(\tau^D)_p(X) = u_p((\theta_{-1})_p X) + u'_p((\omega_0)_p X) + u''_p((\omega_1)_p X) \ ,$$

$X \in T_p(F^2M)$, $p \in F^2M$, where $u'_p \in gl(n, \mathbf{R})^*$, $u''_p \in (S^2(n))^*$ are the elements induced by $u_p \in (\mathbf{R}^n)^*$.

If $\tau = \tau_i \, dx^i$, then with respect to the adapted coframe field we have

$$\tau^D = \tau_i \eta^i + \sum_{j=1}^{n} \tau_i \eta^i_j + \sum_{j,k=1}^{n} \tau_i \eta^i_{jk} \quad .$$

Diagonal lifts of $(1,1)$–tensor fields

Let F be an arbitrary tensor field of type $(1,1)$ on M. For each $p \in F^2M$, we put

$$u_p = (\pi^2_1(p))^{-1} \circ F_x \circ \pi^2_1(p) \quad , \quad x = \pi^2(p) \quad .$$

Then the *diagonal lift* F^D *of* F *to* F^2M is the tensor field of type $(1,1)$ given by

(10.13) $\quad (F^D)_p X = C(u_p((\theta_{-1})_p X)) + \lambda(u'_p((\omega_0)_p X)) + \lambda(u''_p((\omega_1)_p X)) \; ,$

for arbitrary $X \in T_p(F^2M)$, $p \in F^2M$, where $u'_p \in Hom(gl(n,\mathbf{R}), gl(n,\mathbf{R}))$ and $u''_p \in Hom(S^2(n), S^2(n))$ are the elements induced by $u_p \in Hom(\mathbf{R}^n, \mathbf{R}^n)$ according to the definitions in the previous subsection.

With respect to the adapted frame field in F^2M we have

$$F^D = F^h_j \, D_h \otimes \eta^j + \delta^i_j F^h_k \, D^i_h \otimes \eta^k_j + \delta^i_j \delta^k_l F^h_m \, D^h_{jl} \otimes \eta^m_{ik} \quad . \tag{10.14}$$

This definition can be extended in an obvious way for tensor fields of type $(1,s)$, $s \geq 2$. Details are left to the reader.

Proposition 10.7.1

$$\begin{cases} F^D X^H &= (FX)^H \; , \\ F^D(\lambda f) &= \lambda(F^0 f) \; , \\ F^D(\lambda g) &= \lambda(F^0 g) \; , \\ F^D(\lambda A) &= \lambda(F^0 A) \; , \\ F^D(\lambda \alpha) &= \lambda(F^0 \alpha) \; . \end{cases} \tag{10.15}$$

for all vector fields X *on* M, *all* $A \in gl(n,\mathbf{R})$ *and* $\alpha \in S^2(n)$, *and all functions* $f \colon F^2M \to gl(n,\mathbf{R})$ *and* $g \colon F^2M \to S^2(n)$.

Proof. It follows directly from (10.3), (10.8), (10.13) and (10.14), taking into account that $F^0(p) = u_p$, $p \in F^2M$.□

From Proposition 10.7.1, it follows easily

Proposition 10.7.2 *Let F, G be $(1,1)$-tensor fields on M and denote by I the identity tensor field. Then:*

$$(FG)^D = F^D G^D \quad , \qquad I^D = I \ .$$

As a direct consequence of Proposition 10.7.2, we deduce that if $P(t)$ is a polynomial in t, then

$$(P(F))^D = P(F^D) \ .$$

Hence

Corollary 10.7.3 *Let F be a $(1,1)$-tensor field on M. If F defines on M a polynomial structure of rank r and structural polynomial $P(t) = 0$, then F^D defines on F^2M a polynomial structure of rank $r(1+n+(n(n+1)/2))$ and with the same structural polynomial. In particular, if F is an almost complex structure on M, then F^D is an almost complex structure on F^2M.*

Now, let N_{F^D} and N_F be the Nijenhuis tensors of F^D and F respectively. From (10.11) and (10.12) we easily deduce the following identities:

$$\begin{aligned} N_{F^D}(X^H,Y^H) = \{N_F(X,Y)\}^H &- \lambda\{(R(FX,FY) - FR(FX,Y) \\ &\qquad -FR(X,FY) + F^2R(X,Y))^0\} \\ &- 2\lambda\{\Omega_1((FX)^H,(FY)^H) - F^0\Omega_1((FX)^H,Y^H) \\ &\qquad F^0\Omega_1(X^H,(FY)^H) + (F^2)^0\Omega_1(X^H,Y^H)\} \ , \\ N_{F^D}(X^H,\lambda B) &= \lambda\left((\nabla_{FX}F - F\nabla_X F)^0 B\right) \ , \\ N_{F^D}(X^H,\lambda\beta) &= \lambda\left((\nabla_{FX}F - F\nabla_X F)^0 \beta\right) \ , \\ N_{F^D}(\lambda A,\lambda B) &= N_{F^D}(\lambda A,\lambda\beta) = N_{F^D}(\lambda\alpha,\lambda\beta) = 0 \ , \end{aligned}$$

for all vector fields X, Y on M, all $A, B \in gl(n,\mathbf{R})$ and all $\alpha,\beta \in S^2(n)$. Therefore, we have:

Proposition 10.7.4 *The condition*

$$N_{F^D} = 0$$

is equivalent to the conditions

$$\begin{aligned} & N_F = 0 \ , \\ & F\nabla_X F - \nabla_{FX}F = 0 \ , \\ & R(FX,FY) - FR(FX,Y) - FR(X,FY) + F^2R(X,Y) = 0 \ , \\ & \Omega_1((FX)^H,(FY)^H) - F^0\Omega_1((FX)^H,Y^H) \\ & \qquad - F^0\Omega_1(X^H,(FY)^H) + (F^2)^0\Omega_1(X^H,Y^H) = 0 \ . \end{aligned}$$

The last three conditions can be written equivalently as

$$F_i^k \nabla_k F_j^h - F_k^h \nabla_i F_j^k = 0 \ ,$$
$$R_{klm}^h F_j^l F_i^m - R_{kli}^m F_j^l F_m^h - R_{klj}^m F_i^l F_m^h + R_{kji}^l F_l^m F_m^h = 0 \ ,$$
$$R_{kjlm}^h F_s^l F_t^m - R_{kjlt}^i F_s^l F_i^h - R_{kjsm}^i F_i^h F_t^m + R_{kjst}^i F_r^h F_i^r = 0 \ .$$

Diagonal lifts of $(0,2)$–tensor fields

Let G be a $(0,2)$–tensor field on M. For each $p \in F^2M$, we put

$$u_p = G_x \circ (\pi_1^2(p) \times \pi_1^2(p)) \ , \quad x = \pi^2(p) \ .$$

Then the *diagonal lift* G^D *of* G *to* F^2M is the tensor field of type $(0,2)$ on F^2M given by

$$G^D(X,Y) \ = \ u_p((\theta_{-1})_p X, (\theta_{-1})_p Y) + u_p'((\omega_0)_p X, (\omega_0)_p Y) + u_p''((\omega_1)_p X, (\omega_1)_p Y) \ ,$$

where $X, Y \in T_p(F^2M)$, $p \in F^2M$, and with $u_p' \in (gl(n,\mathbf{R}))^* \otimes (gl(n,\mathbf{R}))^*$ and $u_p'' \in (S^2(n))^* \otimes (S^2(n))^*$ are induced by $u_p \in (\mathbf{R}^n)^* \otimes (\mathbf{R}^n)^*$.

If G has local components G_{ij}, then we have

$$G^D = G_{ij}\,\eta^i \otimes \eta^j + \delta^{kl} G_{ij}\,\eta_k^i \otimes \eta_l^j + \delta^{km}\delta^{lr} G_{ij}\,\eta_{kl}^i \otimes \eta_{mr}^j \tag{10.16}$$

in F^2U.

From (10.15) we deduce that if G has constant rank r, then G^D has constant rank $r(1 + n + (n(n+1)/2))$. Thus we have

Proposition 10.7.5 (1): *If G is a Riemannian metric on M, then G^D is a Riemannian metric on F^2M.*

(2): *If G is an almost symplectic form on M, then G^D is an almost symplectic form on F^2M.*

Next, we introduce two new definitions.

Let $A, B \in gl(n,\mathbf{R})$, $\alpha, \beta \in S^2(n)$ and let G be a tensor field of type $(0,2)$ on M with local components G_{ij}. Then

$$G^0(A,B) = \delta^{rs} A_r^k B_s^l x_k^i x_l^j G_{ij} \ , \tag{10.17}$$

and

$$G^0(\alpha,\beta) = \delta^{rt}\delta^{sh} \alpha_{rs}^k \beta_{th}^l x_k^i x_l^j G_{ij} \ , \tag{10.18}$$

are globally defined functions on F^2M where

$$A = A^i_j E^j_i \quad , \quad B = B^i_j E^j_i \quad , \quad \alpha = \alpha^i_{jk} E^{jk}_i \quad , \quad \beta = \beta^i_{jk} E^{jk}_i \quad .$$

One easily obtains the following identities:

$$(10.19) \qquad \begin{cases} G^D(\lambda A, \lambda B) = G^0(A, B) \quad , \\ G^D(\lambda A, \lambda\beta) = G^0(\lambda\beta, \lambda A) \; = \; 0 \quad , \\ G^D(\lambda A, X^H) = G^D(X^H, \lambda A) \\ \qquad\qquad = G^D(\lambda\alpha, X^H) \\ \qquad\qquad = G^D(X^H, \lambda\alpha) = 0 \quad , \\ G^D(\lambda\alpha, \lambda\beta) = G^0(\alpha, \beta) \quad , \\ G^D(X^H, Y^H) = \{G(X, Y)\}^V \quad , \end{cases}$$

where $f^V = f \circ \pi^2$ for any function f on M.

From (10.12), (10.16)–(10.17) and Remark 10.6.1 we deduce the following set of formulas:

$$\begin{aligned} (\mathcal{L}_{\lambda A} G^D)(X^H, \lambda B) &= (\mathcal{L}_{\lambda A} G^D)(\lambda B, X^H) \; = \; 0 \, , \\ (\mathcal{L}_{\lambda A} G^D)(X^H, \lambda\beta) &= (\mathcal{L}_{\lambda A} G^D)(\lambda\beta, X^H) \; = \; 0 \, , \\ (\mathcal{L}_{\lambda A} G^D)(X^H, Y^H) &= 0 \, , \\ (\mathcal{L}_{\lambda A} G^D)(\lambda B, \lambda C) &= G^0(B(A + A^t), C) \, , \\ (\mathcal{L}_{\lambda A} G^D)(\lambda B, \lambda\beta) &= (\mathcal{L}_{\lambda A} G^D)(\lambda\beta, \lambda B) \; = \; 0 \, , \\ (\mathcal{L}_{\lambda A} G^D)(\lambda\alpha, \lambda\beta) &= G^0(\alpha \circ (I, A + A^t), \beta) \\ &\quad + G^0(\alpha, \beta \circ (I, A + A^t)) \, , \\ (\mathcal{L}_{\lambda\alpha} G^D)(X^H, \lambda B) &= (\mathcal{L}_{\lambda\alpha} G^D)(\lambda B, X^H) \; = \; 0 \, , \\ (\mathcal{L}_{\lambda\alpha} G^D)(X^H, \lambda\beta) &= (\mathcal{L}_{\lambda\alpha} G^D)(\lambda\beta, X^H) \; = \; 0 \, , \\ (\mathcal{L}_{\lambda\alpha} G^D)(X^H, Y^H) &= (\mathcal{L}_{\lambda\alpha} G^D)(\lambda A, \lambda B) \; = \; 0 \, , \\ (\mathcal{L}_{\lambda\alpha} G^D)(\lambda A, \lambda\beta) &= (\mathcal{L}_{\lambda\alpha} G^D)(\lambda\beta, \lambda A) \\ &= (\mathcal{L}_{\lambda\alpha} G^D)(\lambda\beta, \lambda\gamma) \; = \; 0 \, , \end{aligned}$$

$$\begin{aligned} (\mathcal{L}_{X^H} G^D)(Y^H, Z^H) &= \{(\mathcal{L}_X G)(Y, Z)\}^V \quad , \\ (\mathcal{L}_{X^H} G^D)(Y^H, \lambda A) &= G^D(\lambda(R(X, Y)^0), \lambda A) \quad , \\ (\mathcal{L}_{X^H} G^D)(\lambda A, Y^H) &= G^D(\lambda A, \lambda(R(X, Y)^0)) \quad , \\ (\mathcal{L}_{X^H} G^D)(Y^H, \lambda\alpha) &= 2G^D(\lambda\Omega_1(X^H, Y^H), \lambda\alpha) \quad , \end{aligned}$$

$$\begin{aligned}(\mathcal{L}_{X^H}G^D)(\lambda\alpha, Y^H) &= 2G^D(\lambda\alpha, \lambda\Omega_1(X^H, Y^H)) \ ,\\ (\mathcal{L}_{X^H}G^D)(\lambda A, \lambda B) &= (\nabla_X G)^0(A, B) \ ,\\ (\mathcal{L}_{X^H}G^D)(\lambda A, \lambda\alpha) &= (\mathcal{L}_{X^H}G^D)(\lambda\alpha, \lambda A) = 0 \ ,\\ (\mathcal{L}_{X^H}G^D)(\lambda\alpha, \lambda\beta) &= (\nabla_X G)^0(\alpha, \beta) \ .\end{aligned}$$

Corollary 10.7.6 *Let G be a Riemannian metric (resp. an almost symplectic form) on M. Then:*

(1): *the fundamental vector field λA is a Killing vector field (resp. an infinitesimal automorphism) of (F^2M, G^D) if and only if $A + A^t = 0$, i.e A is skew-symmetric;*

(2): *the fundamental vector field $\lambda\alpha$ is always a Killing vector field (resp. an infinitesimal automorphism) of (F^2M, G^D).*

Corollary 10.7.7 *The condition $\mathcal{L}_{X^H}G^D = 0$ is equivalent to the conditions*

$$\mathcal{L}_X G = 0 \ , \quad R(X, -) = 0 \ , \quad \nabla_X G = 0 \ , \quad \iota_{X^H}\Omega_1 = 0 \ ,$$

where $R(X, -)$ denotes the tensor field of type $(1,2)$ on M given by

$$R(X, -)(Y, Z) = R(X, Y)Z \ .$$

Now, let us assume that the second order connection Γ induces on M the Levi-Civita connection ∇ of a Riemannian metric G on M. Then we have:

Theorem 10.7.8 *If X^H is a Killing vector field in (F^2M, G^D) then X is a Killing vector field in (M, G). Conversely, suppose that Γ is partially flat (i.e. $\Omega_1 = 0$) and X is a Killing vector field with vanishing second covariant derivative in (M, G); then X^H is a Killing vector field in (F^2M, G^D).*

Proof. We only need to prove the converse. In fact, if X is a Killing vector field in (M, G) then $\mathcal{L}_X G = 0$ and thus $\mathcal{L}_X \nabla = 0$. Therefore $R(X, Y) = -\nabla_Y(\nabla X)$. Moreover, since $(\nabla_Y(\nabla X))(Z) = (\nabla^2 X)(Z, Y)$, we deduce

$$R(X, Y)Z = -(\nabla^2 X)(Z, Y)$$

which ends the proof.□

Next, let G be an almost symplectic form on M. A straightforward computation shows that

$$dG^D(\lambda A, \lambda\beta, \lambda\gamma) = (1/3)\{G^0(\beta \circ (I, A) + \beta \circ (A, I), \gamma) + G^0(\beta, \gamma \circ (I, A), \gamma \circ (A, I))\} \ .$$

Therefore, if $A = I$ we have

$$dG^D(\lambda I, \lambda\beta, \lambda\gamma) = (4/3)G^0(\beta, \gamma) \ .$$

Consequently:

Proposition 10.7.9 *The almost symplectic form G^D is never closed and therefore the almost symplectic manifold (F^2M, G^D) is never symplectic.*

F^2M for an almost Hermitian manifold M

Let us suppose that M is an almost Hermitian manifold with almost complex structure J and Hermitian metric G.

Let Γ be a second order connection on M. We have:

Proposition 10.7.10 *(J^D, G^D) is an almost Hermitian structure on F^2M.*

Now, let Φ be the Kähler form of (M, J, G). A straightforward computation shows the following

Proposition 10.7.11 *The Kähler form of (J^D, G^D) on F^2M is the diagonal lift Φ^D of the Kähler form Φ of (J, G).*

From Propositions 10.7.4 and 10.7.9 it follows:

Theorem 10.7.12
(1): *(J^D, G^D) is never almost Kähler.*
(2): *If (J, G) is Kähler and if its Levi–Civita connection is the linear connection induced from Γ, then (J^D, G^D) is an Hermitian manifold if Γ has zero curvature.*

10.8 Natural prolongations of G–structures

Let $J_p^2\mathbf{R}^n$ be the tangent bundle of p^2–velocities of $\mathbf{R}^n$. Then $J_p^2\mathbf{R}^n$ is a vector space of dimension $N = n + np + np(p+1)/2$. In fact, we define the vector space operations as follows:

$$j^2(\varphi) + j^2(\psi) = j^2(\varphi + \psi) \quad , \quad \lambda j^2(\varphi) = j^2(\lambda\varphi) \quad ,$$

where $j^2(\varphi), j^2(\psi) \in J_p^2\mathbf{R}^n$, $\lambda \in \mathbf{R}$.

Let $Gl(n, \mathbf{R}) \times \mathbf{R}^n \longrightarrow \mathbf{R}^n$ be the natural action of $Gl(n, \mathbf{R})$ on $\mathbf{R}^n$. Then there exists an induced natural action

$$J_p^2 Gl(n, \mathbf{R}) \times J_p^2\mathbf{R}^n \longrightarrow J_p^2\mathbf{R}^n \quad .$$

Thus, each element $j^2(A) \in J_p^2 Gl(n, \mathbf{R})$ defines a linear map

$$j^2(A) : \mathbf{R}^N \longrightarrow \mathbf{R}^N \quad .$$

In fact, $j^2(A)$ is a linear isomorphism with inverse $j^2(A^{-1})$, where

$$\begin{array}{rcl} A^{-1}: \mathbf{R}^p & \longrightarrow & Gl(n,\mathbf{R}) \\ t & \longmapsto & (A(t))^{-1} \end{array}$$

Therefore, there exists a canonical Lie group homomorphism

$$j_p^2: J_p^2 Gl(n,\mathbf{R}) \longrightarrow Gl(N,\mathbf{R}) \ .$$

(In other words, j_p^2 is a linear representation of $J_p^2 Gl(n,\mathbf{R})$ in the vector space $\mathbf{R}^N$).

Imbedding of $J_n^2 FM$ into FF^2M

Take

$FM(M, \pi_M, Gl(n,\mathbf{R}))$: the frame bundle of M ,

$J_n^2 FM(J_n^2 M, \pi_M^{(2)}, J_n^2 Gl(n,\mathbf{R}))$: the induced bundle ,

$FJ_n^2 M(J_n^2 M, \pi_{J_n^2 M}, Gl(N,\mathbf{R}))$: the frame bundle of $J_n^2 M$

(here $N = n + n^2 + (n^2(n+1)/2)$).

Theorem 10.8.1 *There exists a canonical injective morphism of principal bundles*

$$j_M^2: J_n^2 FM \longrightarrow FJ_n^2 M$$

over the identity of $J_n^2 M$, with associated Lie group homomorphism

$$j_n^2: J_n^2 Gl(n,\mathbf{R}) \longrightarrow Gl(N,\mathbf{R}) \ .$$

The proof is similar to that of Theorem 2.1.2.

Now, let $J_n^2 F^2 M_{|F^2M}$ be the restriction of $J_n^2 F^2 M$ to the open submanifold F^2M of $J_n^2 M$. Note that the restriction $FJ_n^2 M_{|F^2M}$ is canonically isomorphic to the frame bundle FF^2M of F^2M. Then from Theorem 10.8.1, we deduce:

Theorem 10.8.2 *j_M^2 induces a bundle morphism of $J_n^2 FM_{|F^2M}$ into FF^2M over the identity of F^2M and with associated Lie group homomorphism j_n^2.*

Let $G \subset Gl(n,\mathbf{R})$ be a Lie subgroup and put

$$G^{(2)} = j_n^2(J_n^2 G) \ .$$

Then $G^{(2)}$ is a Lie subgroup of $Gl(N,\mathbf{R})$ isomorphic to $J_n^2 G$. Let $P(M,\pi,G)$ be a G-structure on M. Then:

Theorem 10.8.3 *If M has a G-structure P, then F^2M has a canonical $G^{(2)}$-structure $P^{(2)}$.*

Proof. Put $P^{(2)} = j_M^2(J_n^2 P_{|F^2M})$. □

The $G^{(2)}$-structure $P^{(2)}$ will be called *the natural prolongation to F^2M of the G-structure P on M.*

The following theorem can be proved similarly to that in the case of the prolongation of a G-structure to the frame bundle FM (Section 2.2).

Theorem 10.8.4 *P is integrable if and only if its natural prolongation $P^{(2)}$ is integrable.*

Applications

Next we shall describe, as an application, the prolongation of some classical G-structures.

Let P be a G-structure on M, (X, x^i) a local chart and $\phi: U \longrightarrow P$ a section over U. Then ϕ defines a local field of frames $\{X_1, \ldots, X_n\}$ adapted to P (cf. 2.4) and given by

$$X_i = \phi_i^j \frac{\partial}{\partial x^j} \quad .$$

Hence, the local field of coframes $\{\theta^1, \ldots, \theta^n\}$ dual to $\{X_1, \ldots, X_n\}$ is given by

$$\theta^j = \psi_i^j \, dx^i \quad ,$$

where (ψ_i^j) denotes the inverse matrix of (ϕ_j^i). Moreover, ϕ induces a section

$$\overline{\phi} = \phi^{(2)}{}_{|F^2U}: F^2U \longrightarrow FF^2M \quad .$$

Linear endomorphisms

Let $\rho: Gl(n, \mathbf{R}) \longrightarrow Aut(\mathbf{R}^n)$ be the canonical representation of $Gl(n, \mathbf{R})$ in $\mathbf{R}^n$, $u \in Hom(\mathbf{R}^n, \mathbf{R}^n)$ an arbitrary element and G_u the isotropy group of u with respect to ρ.

Let $u^{(2)} = j_n^2 u \in Hom(\mathbf{R}^N, \mathbf{R}^N)$ be the induced map. Then, a straightforward computation yields the following

Proposition 10.8.5 *Let $G_{u^{(2)}}$ be the isotropy group of $u^{(2)}$ with respect to the canonical representation of $Gl(N, \mathbf{R})$ into $\mathbf{R}^N$ and denote $(G_u)^{(2)} = j_n^2(J_n^2 G_u)$. Then*

$$(G_u)^{(2)} \subset G_{u^{(2)}} \quad .$$

Theorem 10.8.6 *If a manifold M admits a G_u–structure, then F^2M admits a $G_{u^{(2)}}$–structure. Moreover, if the G_u–structure on M is integrable, then also the induced $G_{u^{(2)}}$–structure on F^2M is integrable.*

Proof. Theorem 10.8.3 implies that F^2M admits a $(G_u)^{(2)}$–structure which, by Proposition 10.8.5, can be extended to a $G_{u^{(2)}}$–structure. If the G–structure on M is integrable, then also the $G_{u^{(2)}}$–structure on F^2M is integrable, by Theorem 10.8.4.□

Now, let P be a G_u–structure on M and let F be the $(1,1)$–tensor field associated to P. Then, F is locally given by

$$F = F^i_j \frac{\partial}{\partial x^i} \otimes dx^j = u^i_j X_i \otimes \theta^j \quad .$$

A straightforward computation shows that the $(1,1)$–tensor field $F^{(2)}$ on F^2M associated to $P^{(2)}$ is precisely the complete lift F^C of F to F^2M defined by J.Gancarzewicz (see [35]). Therefore,

Theorem 10.8.7 *The complete lift F^C of F to F^2M defines the $G_{u^{(2)}}$–structure $P^{(2)}$ on F^2M given in* Theorem 10.8.6.

Bilinear forms

By a similar procedure, we can show how the different definitions of lifts of tensor fields of type $(0,2)$ on M to F^2M given by J. Gancarzewicz in [36] may be obtained through a similar construction. It suffices to choose the appropriate prolongation to $(\mathbf{R}^N)^* \otimes (\mathbf{R}^N)^*$ of the "model" element $u \in (\mathbf{R}^n)^* \otimes (\mathbf{R}^n)^*$. We refer the reader to [54] for more details.

10.9 Diagonal prolongation of G–structures

Let Γ be a second order connection on M, and let $G \subset Gl(n,\mathbf{R})$ a Lie subgroup. Define an injective Lie group homomorphism by

$$\begin{array}{rcl} i\colon G & \longrightarrow & J^2_n G \\ g & \longmapsto & i(g) = j^2(\tilde{g}) \end{array}$$

where $\tilde{g}\colon \mathbf{R}^n \longrightarrow G$ is the constant map

$$\tilde{g}(t) \;=\; g \quad , \quad t \in \mathbf{R}^n \quad .$$

Now, we put

$$G_0 \;=\; (j^2_n \circ i)(G) \quad ;$$

then G_0 is a Lie subgroup of $Gl(N, \mathbf{R})$.

Let $P(M, \pi, G)$ be a G-structure on M and $\phi: U \to G$ a local section of P over a coordinate neighbourhood (U, x^i):

$$\phi(x^i) = (x^i, \phi^j_i(x)) \quad .$$

Then ϕ defines a local field of frames adapted to P and given by

$$X_i = \phi^j_i \frac{\partial}{\partial x^i} \quad .$$

We can associate to ϕ a section

$$\phi_\Gamma: F^2U \longrightarrow F(F^2M)$$

defined by

$$\phi_\Gamma(j^2\phi) = \{X^H_i, X^j_i, X^{jk}_i\} \quad , \quad 1 \leq i,j,k \leq n \quad , \quad j \leq k \quad ,$$

where

$$\begin{aligned} X^H_i &= (\phi^j_i \frac{\partial}{\partial x^j})^H = \phi^j_i D_j \quad , \\ X^j_i &= \sum_k \phi^k_i D^k_j \quad , \\ X^{jk}_i &= \sum_k \phi^k_i D^k_{jk} \quad , \end{aligned}$$

where $\{D_i, D^i_j, D^i_{jk}\}$ are the adapted frame fields on F^2U determined by Γ.

Let $\phi': U' \longrightarrow P$ be another section, with (U', x'^i) another local coordinate system intersecting (U, x^i). Then there exists a differentiable map

$$\mathsf{a}: U \cap U' \longrightarrow G$$

such that

$$\phi'(x) = \phi(x)\mathsf{a}(x) \quad , \quad x \in U \cap U' \quad .$$

Then, if ϕ'_Γ is the section associated to ϕ', there exists a differentiable map

$$\mathsf{a}_\Gamma: F^2U \cap F^2U' \longrightarrow Gl(N, \mathbf{R})$$

such that

$$\phi'_\Gamma(p) = \phi_\Gamma(p)\mathsf{a}_\Gamma(p) \quad , \quad p \in F^2U \cap F^2U' = F^2(U \cap U') \quad .$$

Now, a simple computation shows that

$$\mathsf{a}_\Gamma(p) = (j^2_n \circ i)(\mathsf{a}(x)) \quad , \quad x = \pi^2(p) \quad .$$

Therefore, we have

Theorem 10.9.1 *Let Γ be a second order connection on M. If M has a G-structure P, then F^2M has a G_0-structure $\tilde{P}_0$ induced by P and Γ.*

The G_0-structure $\tilde{P}_0$ will be called *diagonal prolongation of the G-structure P on M to the frame bundle F^2M with respect to Γ.*

Applications

Linear endomorphisms

Let P be a G_u-structure on M, $u \in Hom(\mathbf{R}^n, \mathbf{R}^n)$, F the $(1,1)$-tensor field on M induced by P, and Γ a second order connection on M.

Then, it can be easily proved that the diagonal lift F^D of F to F^2M with respect to Γ defines a $G_{u(2)}$-structure $\tilde{P}'$ on F^2M induced by the diagonal prolongation $\tilde{P}_0$ of P to F^2M with respect to Γ. (See [57] for the details.)

Bilinear forms

Take $u \in (\mathbf{R}^n)^* \otimes (\mathbf{R}^n)^*$ and G_u the isotropy group of u with respect to the natural representation of $Gl(n, \mathbf{R})$ in $\mathbf{R}^n$.

Define an element $\tilde{u} \in (\mathbf{R}^N)^* \otimes (\mathbf{R}^N)^*$ as that whose matrix representation with respect to the canonical basis of $\mathbf{R}^N$ is

$$\tilde{u} = \begin{pmatrix} (u_{ij}) & \cdots & 0 \\ \vdots & \ddots & \vdots \\ 0 & \cdots & (u_{ij}) \end{pmatrix} ,$$

(u_{ij}) being the matrix representation of u with respect to the canonical basis of $\mathbf{R}^n$.

Then $(G_u)_0 \subset G_{\tilde{u}}$, where $G_{\tilde{u}}$ the isotropy group of $\tilde{u}$ with respect to the canonical representation of $Gl(N, \mathbf{R})$ in $\mathbf{R}^N$.

Therefore, if P is a G_u-structure on M and Γ a second order connection on M, the diagonal prolongation $\tilde{P}_0$ of P to F^2M with respect to Γ induces a $G_{\tilde{u}}$-structure $\tilde{P}'$ on F^2M.

Now, if G is a $(0,2)$-tensor field on M associated to P, then the $(0,2)$-tensor field $\tilde{G}$ on F^2M associated to $\tilde{P}'$ is precisely the diagonal prolongation G^D of G to F^2M with respect to Γ.

Bibliography

[1] **S–I. Amari:** *Differential geometric methods in statistics.* Lect. Notes in Statistics **28** (1985), Springer–Verlag, New York.

[2] **K. Aso:** Note on some properties of the sectional curvature of the tangent bundle. *Yokohama Math. J.* **29**, 1–5 (1980).

[3] **O.E. Barndorff–Nielsen, D.R. Cox, N. Reid:** Differential Geometry in Statistical Theory. *Internat. Statist. Rev.* **54**, 1, 83–96 (1986).

[4] **J.K. Beem, P.E. Ehrlich:** *Global Lorentzian Geometry.* Marcel Dekker Inc, New York 1981.

[5] **J.K. Beem, P.E. Ehrlich:** Geodesic completeness and stability. *Math. Proc. Camb. Phil. Soc.* **102**, 319–328 (1987).

[6] **D.E. Blair:** *Contact Manifolds in Riemannian Geometry.* Lect. Notes in Math. **509** (1976), Springer–Verlag, Berlin.

[7] **A. Bonome, R. Castro, L.M. Hervella:** Almost complex structure in the frame bundle of an almost contact metric manifold. *Math. Z.* **193**, 431–440 (1986).

[8] **D. Canarutto, C.T.J. Dodson:** On the bundle of principal connections and the stability of b–incompleteness of manifolds. *Math. Proc. Camb. Phil. Soc.* **98**, 51–59 (1985).

[9] **N.N. Chentsov:** *Statistical Decision Rules and Optimal Inference* (Russian, Nauka Moscow 1972). English version in *Translations of Mathematical Monographs* vol. **53**, Amer. Math. Soc., Providence, Rhode Island 1982.

[10] **S.S. Chern:** *Topics in Differential Geometry*, Inst. for Adv. Study, Princeton 1951.

[11] **L.A. Cordero, P.M. Gadea:** On the integrability conditions of a structure ϕ satisfying $\phi^4 \pm \phi^2 = 0$. *Tensor N.S.* **28**, 78–82 (1974).

[12] **L.A. Cordero, M. de León:** Lifts of tensor fields to the frame bundle. *Rend. Circ. Mat. Palermo* **32**, 236–271 (1983).

[13] **L.A. Cordero, M. de León:** Prolongation of linear connections to the frame bundle. *Bull. Australian Math. Soc.* **28**, 367–381 (1983).

[14] **L.A. Cordero, M. de León:** Tensor fields and connections on cross–sections in the frame bundle of a parallelizable manifold. *Riv. Mat. Univ. Parma* **(4) 9**, 433–445 (1983).

[15] **L.A. Cordero, M. de León:** Horizontal lifts of connections to the frame bundle. *Bollettino Unione Mat. Italiana* **(6) 3–B**, 223–240 (1984).

[16] **L.A. Cordero, M. de León:** Prolongation of vector–valued differential forms to the frame bundle. *J. Korean Math. Soc.* **21**, 183–196 (1984).

[17] **L.A. Cordero, M. de León:** On the differential geometry of the frame bundle. *Rev. Roumaine Math. Pures Appl.* **31**, 9–27 (1986).

[18] **L.A. Cordero, M. de León:** Prolongations of G–structures to the frame bundle. *Ann. Mat. Pura Appl.* **(IV) 143**, 123–141 (1986).

[19] **L.A. Cordero, M. de León:** On the curvature of the induced Riemannian metric on the frame bundle of a Riemannian manifold. *J. Math. Pures Appl.* **65**, 81–91 (1986).

[20] **L.A. Cordero, M. de León:** f–structures on the frame bundle of a Riemannian manifold. *Riv. Mat. Univ. Parma* (4) **12**, 257–262 (1986).

[21] **L. Del Riego, C.T.J. Dodson:** Sprays, universality and stability. *Math. Proc. Camb. Phil. Soc.* **103**, 3, 515–534 (1988).

[22] **C.T.J. Dodson:** Space–time edge geometry. *Internat. J. Theor. Phys.* **17**, 389–504 (1978).

[23] **C.T.J. Dodson:** *Categories, bundles and spacetime topology*. 2nd Edition D. Reidel, Dordrecht 1988.

[24] **C.T.J. Dodson:** Systems of connections for parametric models. In Proceedings of the *Workshop on Geometrization of Statistical Theory*, ed. C.T.J. Dodson, Lancaster 29–31 October 1987, ULDM Publications, University of Lancaster 1987, 153–169.

[25] **C.T.J. Dodson, M. Modugno:** Connections over connections and universal calculus. *Proc. VI Convegno Nazionale de Relatività Generale e Física della Gravitazione*, Florence 10–13 October 1984, in press.

[26] **C.T.J. Dodson, L.J. Sulley:** The b-boundary of S^1 with constant connection. *Lett. Math. Phys.* **1**, 301-307 (1977).

[27] **C.T.J. Dodson, E. Vázquez-Abal:** Tangent and frame bundle harmonic lifts. *Mathematicheskie Zametki*, in press.

[28] **P. Dombrowski:** On the geometry of the tangent bundles. *J. reine angew. Math.* **210**, 73-88 (1962).

[29] **J. Eells, L. Lemaire:** A report on harmonic maps. *Bull. London Math. Soc.* **10**, 1-68 (1978).

[30] **J. Eells, L. Lemaire:** *Selected topics in harmonic maps.* Conference Series in Math. n^0 **50**, Amer. Math. Soc. 1983.

[31] **M. Fernández, M. de León:** Some properties of the holomorphic sectional curvature of the tangent bundle. *Rend. Sem. Fac. Sc. Univ. Cagliari* **56**, 11-19 (1986).

[32] **A. Fujimoto:** *Theory of G-structures. Publ. Study Group of Geometry* **1**, Tokyo Univ., Tokyo 1972.

[33] **J. Gancarzewicz:** Connections of order two. *Zeszyty Nauk. Univ. Jagiellonski, Prace Mat.* **19**, 121-136 (1977).

[34] **J. Gancarzewicz:** Connections of order r. *Ann. Pol. Math.* **34**, 70-83 (1977).

[35] **J. Gancarzewicz:** Complete lifts of tensor fields of type $(1, k)$ to natural bundles. *Zeszyty Nauk. Univ. Jagiellonski, Prace Mat.* **23**, 51-84 (1982).

[36] **J. Gancarzewicz:** Liftings of functions and vector fields to natural bundles. *Dissertationes Mathematicae* **XII**, 1983.

[37] **P.L. García:** Connections and 1-jet fiber bundles. *Rend. Sem. Mat. Univ. Padova* **47**, 227-242 (1972).

[38] **S.I. Goldberg, N.C. Petridis:** Differential solutions of algebraic equations on manifolds. *Kōdai Math. Sem. Rep.* **25**, 111-129 (1973).

[39] **M.J. Gotay, J.A. Isenberg:** Geometric quantization and gravitational collapse. *Phys. Rev.* **D 22**, 235-260 (1980).

[40] **A. Gray:** Riemannian almost product manifolds and submersions. *J. Math. Mech.* **16**, 715-737 (1967).

[41] **A. Gray:** Curvature identities for Hermitian and almost Hermitian manifolds. *Tohôku Math. J.* **28**, 601–612 (1976).

[42] **S.W. Hawking, G.F.R. Ellis:** *The Large Scale Structure of Space–Time.* Cambridge Univ. Press, Cambridge 1974.

[43] **S. Ishihara, K. Yano:** On integrability conditions of a structure f satisfying $f^3 + f = 0$. *Quart. J. Math. Oxford* **15**, 217–222 (1964).

[44] **S. Kobayashi:** Theory of connections. *Ann. Mat. Pura Appl.* **43**, 119–194 (1957).

[45] **S. Kobayashi:** Canonical forms on frame bundles of higher order contact. *Proc. Symp. Pure Math.* **3**, 186–193 (1961).

[46] **S. Kobayashi:** *Transformation Groups in Differential Geometry.* Springer, Berlin 1972.

[47] **S. Kobayashi, K. Nomizu:** *Foundations of Differential Geometry, vol. I.* Interscience Publ., New York 1963.

[48] **S. Kobayashi, K. Nomizu:** *Foundations of Differential Geometry, vol. II.* Interscience Publ., New York 1969.

[49] **O. Kowalski:** Curvature of the induced Riemannian metric on the tangent bundle of a Riemannian manifold. *J. reine angew. Math.* **250**, 124–129 (1971).

[50] **J. Lehmann–Lejeune:** Integrabilité des G–structures définies par une 1–forme 0–deformable à valeurs dans le fibré tangent. *Ann. Inst. Fourier* **16**, 329–287 (1966).

[51] **M. de León, M. Salgado:** A characterization of geodesics of second order. *Bol. Acad. Galega de Ciencias* **2**, 103–106 (1983).

[52] **M. de León, M. Salgado:** G–structures on the frame bundle of second order. *Riv. Mat. Univ. Parma* (4) **11**, 161–179 (1985).

[53] **M. de León, M. Salgado:** Lifts of derivations to the tangent bundle of p^r–velocities. *J. Korean Math. Soc.* **23**, 135–140 (1986).

[54] **M. de León, M. Salgado:** Diagonal lifts of tensor fields to the frame bundle of second order. *Acta Sci. Math.* **50**, 67–86 (1986).

[55] **M. de León, M. Salgado:** Levantamientos completos y horizontales de campos de vectores a fibrados naturales. *Act. IX Jorn. Mat. Hispano–Lusas, Univ. Extremadura*, vol. **II**, 223–230 (1986).

[56] **M. de León, M. Salgado:** Lifts of derivations to the frame bundle. *Czechoslovak Math. J.* **23**, 135–140 (1986).

[57] **M. de León, M. Salgado:** Diagonal prolongations of G–structures to the frame bundle of second order. *Ann. Univ. Bucuresti* **34**, 40–51 (1987).

[58] **M. de León, M. Salgado:** Tensor fields and connections on cross–sections in the frame bundle of second order. *Publ. Inst. Mathematique* **43 (57)**, 83–87 (1988).

[59] **M. de León, M. Salgado:** Prolongation of G–structures to the frame bundle of second order. *Publ. Mathematicae Debrecen*, in press.

[60] **P. Libermann:** Calcul tensoriel et connexions d'ordre supérieur. *An. Acad. Brasil Ciencias* **37**, 17–29 (1985).

[61] **L. Mangiarotti, M. Modugno:** Fibred spaces, jet spaces and connections for field theories. *Proc. Internat. Meeting Geometry and Physics*, Firenze 12-15 Oct. 1982, 135–165, Pitágora Editrice, Bologna 1983.

[62] **K.B. Marathe:** A condition for paracompactness for manifolds. *J. Diff. Geom.* **7**, 571–573 (1972).

[63] **M. Modugno:** An introduction to systems of connections. Preprint, Ist. Mat. Appl. "G. Sansone", Florence 1986.

[64] **K.P. Mok:** On the differential geometry of frame bundles of Riemannian manifolds. *J. reine angew. Math.* **302**, 16–31 (1976).

[65] **K.P. Mok:** Complete lift of tensor fields and connections to the frame bundle. *Proc. London Math. Soc.* **(3) 32**, 72–88 (1979).

[66] **A. Morimoto:** On normal almost contact structures. *J. Math. Soc. Japan* **15**, 420–436 (1963).

[67] **A. Morimoto:** Prolongation of G–structures to tangent bundles. *Nagoya Math. J.* **32**, 67–108 (1968).

[68] **A. Morimoto:** *Prolongation of Geometric Structures.* Math. Inst. Nagoya Univ., Nagoya 1969.

[69] **A. Morimoto:** Prolongation of connections to tangential fibre bundles of higher order. *Nagoya Math. J.* **40**, 85–97 (1970).

[70] **M.S. Narasimhan, S. Ramanan:** Existence of universal connections, I. *Amer. J. Math.* **83**, 563–572 (1961).

[71] **M.S. Narasimhan, S. Ramanan:** Existence of universal connections, II. *Amer. J. Math.* **85**, 223–231 (1963).

[72] **K. Nomizu:** *Lie groups and Differential Geometry*, Publ. Math. Soc. Japan, # **2**, 1956.

[73] **T. Okubo:** On the differential geometry of frame bundles $F(X_n)$, $n = 2m$. *Memoir Defense Acad.* **5**, 1–17 (1965).

[74] **T. Okubo:** On the differential geometry of frames bundles. *Ann. Mat. Pura Appl.* **72**, 29–44 (1966).

[75] **J.A. Oubiña:** On almost complex structures on the semidirect products of almost contact Lie algebras. *Tensor N.S.* **41**, 111–115 (1984).

[76] **A. Roux:** Jet et connexions. *Publ. Dep. de Mathématiques*, Lyon, 1975.

[77] **M. Salgado:** Sobre la geometría diferencial del fibrado de referencias de orden 2. *Pub. Dep. Geometría y Topología* **63**, Univ. Santiago, Spain 1984.

[78] **A. Sanini:** Applicazioni armoniche tra i fibrati tangenti di varietá riemanniane. *Bollettino U.M.I.* **2–A**, 55–63 (1983).

[79] **S. Sasaki:** On the differential geometry of tangent bundles of Riemannian manifolds. *Tohôku Math. J.* **10**, 338–354 (1958).

[80] **B.G. Schmidt:** A new definition of singular points in general relativity. *Gen. Relativity Gravitation* **1**, 269–280 (1971).

[81] **R.T. Smith:** Harmonic mappings of spheres. *Amer. J. Math.* **97** (1), 364–385 (1975).

[82] **S. Tanno:** Almost complex structures in bundle spaces over almost contact manifolds. *J. Math. Soc. Japan* **17**, 167–186 (1965).

[83] **J.M. Terrier:** Linear connections and almost complex structures. *Proc. Amer. Math. Soc.* **49**, 59–65 (1975).

[84] **V.C. Vohra, K.D. Singh:** Some structures on an f-structure manifold. *Ann. Polon. Math* **27**, 85–91 (1972).

[85] **P.G. Walczak:** Polynomial structures on principal fiber bundles. *Colloquium Math.* **35**, 73–81 (1976).

[86] **Y.C. Wong:** Recurrent tensors on a linearly connected differentiable manifold. *Trans. Amer. Math. Soc.* **99**, 325–341 (1961).

[87] **K. Yano, S. Ishihara:** Horizontal lifts of tensor fields and connections to tangent bundles. *J. Math. Mech.* **16**, 1015–1030 (1967).

[88] **K. Yano, S. Ishihara:** *Tangent and Cotangent Bundles. Differential Geometry*. Marcel Dekker Inc., New York 1973.

[89] **K. Yano, S. Kobayashi:** Prolongations of tensor fields and connections to tangent bundles. *J. Math. Soc. Japan* **18**, 194–210 (1966).

Index